BODY WORKS

PHYSICS AND CHEMISTRY
FOR NURSES

BODY WORKS

PHYSICS AND CHEMISTRY
FOR NURSES

PAUL STRUBE

UNIVERSITY OF SOUTH AUSTRALIA

PRENTICE HALL HEALTH

Pearson Education Australia
Unit 4, Level 2
14 Aquatic Drive
Frenchs Forest NSW 2086

www.pearsoned.com.au

Acquisitions Editor: Nicole Meehan
Senior Project Editor: Kathryn Fairfax
Editorial Coordinator: Jill Gillies
Copy Editor: Joy Window
Proofreader: Tom Flanagan
Cover and internal design by Liz Seymour of Seymour Designs
Cover photograph from Getty Images
Typeset by Midland Typesetters, Maryborough, Vic.

Printed in Malaysia, CLP

5 07 06 05

National Library of Australia
Cataloguing-in-Publication Data

Strube, Paul.
Body works: physics and chemistry for nurses.

   2nd ed.
   Bibliography.
   Includes index.
   ISBN 1 74009 834 X.

   1. Physics. 2. Chemistry. 3. Nursing. 4. Nursing—Case
   studies. 5. Biophysics. 6. Biochemistry. I. Title.

530.024613

PRENTICE
HALL
HEALTH
An imprint of Pearson Education Australia

CONTENTS

PREFACE

Body Works has been substantially revised for this edition. There have been major structural and content changes to Chapters 6–8, while the text as a whole has been updated and hopefully made even more reader-friendly. There is greater emphasis throughout the text on clinical notes and calculation problems. The chemistry content has been increased and two appendices have been added to deal with mathematics and data handling.

I am grateful to all those students and colleagues who have provided helpful comments and advice. This edition is the better for your help, though I am sure that I have not been able to satisfy everyone's wishes. I hope that your comments and criticism will continue, along with your interest in this text.

This edition continues the emphasis in the first edition, on creating a text that engages the reader, provides important information clearly, and places that information in a relevant nursing context.

TO THE READERS

Body Works has been written with one important purpose in mind: to provide you with the basic physical science you will be required to know as a health professional at the beginning of your career. For that reason, I have been deliberately selective in the amount of science covered in the text, and in the depth to which I have treated it. The view I have taken is that the physical sciences act to support the human anatomy and physiology which lie at the heart of the science required for nursing and allied health professions.

There are several features of this text that I would like to draw your attention to before you start reading it.

The case study approach

Each chapter is introduced by a small case study based on a hospital or clinical situation. These case studies are there to provide a realistic context for the science discussed in that chapter, for I know very well that beginning nursing students are often not yet aware of the variety of settings in which health professionals are called upon to recall and apply scientific ideas. Also, learning is more likely to occur when students find new information in a context where they can see it actually applied. That is the intention of these case studies.

The language of the text

I have made every effort to write this text as though you are present and we are discussing the science together. Many otherwise excellent textbooks are made less accessible to students because their authors assume that the technical language they use is already familiar to the reader. I have started with the assumption that simple, conversational language is perfectly suitable for introducing even the most complex ideas to any reader. It is my hope that the ideas and concepts presented in this text are not clouded behind a difficult and remote style of writing.

I would welcome hearing your views on how this text can be made more useful to you. I wish you all the best with your studies, and in your chosen profession.

TO THE LECTURERS

Body Works is written for the beginning student, and provides the basic physical science required of a beginning nurse at RN or EN level. It assumes that many of its readers will have little or no substantial physical science in their backgrounds; accordingly, the use of mathematics here is limited.

The range of content and depth of treatment have been chosen to provide students with a conceptually clear overview of major concepts, without attempting to provide all the underlying scientific principles appropriate for science students. The depth of coverage will give a beginning nurse an accept-able, practical, working knowledge. It is hoped that the case studies provide opportunities for fuller discussion of the application and relevance of physical science in the nursing curriculum.

Body Works is accompanied by an instructor's manual containing the diagrams in a form suitable for the creation of overhead transparencies, and some suggested laboratory practicals. It is anticipated that the text will soon be supported by a companion website.

PAUL STRUBE
UNIVERSITY OF SOUTH AUSTRALIA

ACKNOWLEDGMENTS

I would like to thank Nicole Meehan of Pearson Education Australia for her support and efforts in bringing this second edition to fruition. To the reviewers who gave so generously of their time and knowledge, my grateful thanks. I would like to dedicate this edition to my wife and children, for their understanding and support.

THE METRIC SYSTEM AND MEASUREMENT

Chapter outline

1.1 **Units and measurements.** Your objective: to understand the need for consistent units and the importance of careful measurements.

1.2 **The SI system.** Your objectives: to be able to recognise the commonly used measurements in nursing; to see the relationship between standards used in measurement and human convenience; to know the seven basic units and their symbols; and to understand how derived units are put together.

1.3 **Unit prefixes.** Your objective: to be able to use appropriately the SI system prefixes—kilo, milli, centi and others.

1.4 **Simple metric conversions.** Your objectives: to be able to perform simple conversions, such as changing large units to small and small to large; to understand and use simple powers of 10 notation; and to practise both worked and unworked examples of simple drug calculations.

INTRODUCTION

This chapter contains a discussion of the units of measurement and some basic mathematical skills that nurses use daily. Much of the information presented here will then be used throughout the rest of this book. As well, there are examples of drug calculation problems. With regular practice these calculations will, I hope, become easy, if not automatic. There are more practice problems for you to try in Appendix 1 at the end of this book.

Case Study: Mr Warren D.

It's your first day of clinical placement, as a nursing student, on the Accident and Emergency Ward. An elderly male, Mr Warren D., has just been admitted. He is unconscious, not bleeding and shows no obvious physical damage such as a broken limb. A series of routine measurements and clinical tests are done. Later in the day the Clinical Nurse Consultant (CNC) takes the time to show you the results.

Height:	178.5 cm
Weight:	109 kg
Temperature on admission:	39.7°C
Blood glucose level:	5 mmol/L
Blood pressure:	120 mm Hg/90 mm Hg
Cardiac output:	4.8 L/min
Red blood cell count:	1 000 000/mm^3

She then asks you two questions. Which of these readings, if any, show cause for alarm? What clues do they provide about Mr D.'s condition?

1.1 UNITS AND MEASUREMENTS

What she has given you is a list of **measurements**. They have two parts—a number, such as 178.5, and a unit, such as centimetres (cm). Both are necessary. Imagine you were told that a client had lost some blood and you asked 'How much?' If the answer was 'Three!', you would still very much like to know if it was 3 millilitres or 3 litres!

Some of the measurements listed above are, perhaps, already familiar to you. For example, you probably realise that Mr D.'s height, weight and temperature do not show any cause for alarm. They are within normal ranges for people who are healthy, even though there is some slight evidence of fever. The idea of a **normal range** is the first reason for learning about measurement and is quite simple. There is no single, exact glucose level or blood pressure that indicates health. These things vary in people, though not usually by very much, in the same way that their height and weight differ. Also, there are normal changes in these values as we age; our normal body temperature, for example, may be slightly higher when we are infants than when we are elderly. The normal range consists of those values of the measurement that reflect individual variations; a temperature slightly away from 37°C is still within the normal range of healthy body temperatures. It is when the measured value falls outside the expected normal range for that measurement that the nurse is alerted to possible trouble.

If all the results of the above tests are not already familiar to you, don't be concerned; in the course of your training you will come to know the meaning and significance of each of these measurements. For example, a skilled nurse would realise that Mr D. has a very low red blood cell count, evidence of anaemia, and this knowledge would help her or him in providing appropriate care.

(Note: in this case, 1 000 000/mm³ means 1 000 000 red blood cells were found in every cubic millimetre of blood. This is the same idea as cm³ or cc, which means cubic centimetres.)

Measurement
A number plus a unit. Measurements are a comparison of one aspect of an object, such as its length, with an agreed upon standard; in this case, the metre.

Normal range
Those values for any measurement that indicate the expected, usual variation between individuals. Measurements whose values fall outside this normal range are taken to be indicators of an abnormal condition. How wide or narrow the normal range for any measurement should be is not always easy to determine, however. Tables of normal values are published by relevant authorities.

Calculation practice

If the normal red blood cell count is 5 000 000/mm³, what percentage of this is Mr D.'s count? (**Answer:** 20%)

A second reason for quickly becoming familiar with commonly used measurements is to prevent error. The following example shows this clearly. A nurse noticed that a postpartum patient with a relatively minor infection was scheduled to receive an unusually high dose of penicillin: 6.5 million units of IV penicillin G every 4 hours, or 39 000 000 units a day. Thinking that another nurse had transcribed the order incorrectly, she looked at the original on the patient's chart and discovered a dangerous misinterpretation. The doctor's sloppy handwriting made the letter 'G' after penicillin look like a 6 and he had written the 500 000 units as '.5 million units', instead of '0.5 million units'. So the order did look like 6.5 million units.

The first nurse should have realised that the apparent dose was much too high. Although it occasionally causes a serious allergic reaction, penicillin G is a very safe drug. But because each million units also contains 1.7 mEq of potassium, a mistake in the administration rate or the dosage could be fatal for a child or for an adult with a serious illness. No matter how rare such misinterpretations may be, your knowledge of measurements and units is a vital professional responsibility.

1.2 THE SI SYSTEM

Modern nursing is, of course, both scientific and international. Therefore the units used for measurements are now based on the international, scientific system known as SI, or Système International d' Unités. It is also known as a metric system, because one of its main units is the metre. Scientists use it everywhere and only a few countries, such as the United States, do not also use it for everyday measurements. One reason it is popular is because it's based on multiples of 10 and easy to use.

The **SI system** has only seven basic types of measurements. Each has its own unit, which has been internationally agreed upon by all those who use that unit. The seven basic types are:

SI System
SI stands for System International, its formal (French) title is Système International d'Unités. The seven standards of comparison in this system are mainly based on the behaviour of atoms.
It is a decimal system, based on powers of 10 and it is also referred to as a metric system, based on the metre.

- length, measured in metres (m)
- mass, measured in kilograms (kg)
- time, measured in seconds (s)
- temperature, measured in kelvins (K) (though degrees Celsius (°C) is more commonly used in practice)
- electric current, measured in amperes (A)
- luminous intensity, measured in candelas (cd)
- quantity of substances, measured in moles (mol)

1.3 UNIT PREFIXES

There are two things to say about this list. First, we often use prefixes in front of these units to show when smaller or larger subunits are being used. For example, long distances can be measured in kilometres and small masses can be measured in milligrams. Those prefixes are given in Table 1.1.

Second, there are many more than seven types of measurements, of course. Blood pressure and surface area are two obvious examples. The units used for all these other measurements, however, are simply combinations of the seven basic ones. Area, for example, is often measured in m × m, or m², or as cm × cm, or cm². Pressure is measured in units called pascals (Pa), which are a combination of kilograms, metres and seconds. These combinations are referred to as **derived units** and they will be pointed out to you as they appear in the text. Table 1.2 presents some of the more common derived units.·

You may be unfamiliar with the notation used for the definition of units, such as kg.m².s⁻². First, consider the m². The ² is called a positive exponent. It is short

Derived units
Units that are combinations of two or more of the seven basic SI units.

TABLE 1.1 METRIC SYSTEM PREFIXES

Name	Prefix	Symbol	Number	Power of ten
Ten	deka	da	10	10^1
Hundreds	hecto	h	100	10^2
Thousands	kilo	k	1000	10^3
Millions	mega	M	1 000 000	10^6
Tenths	deci	d	0.1	10^{-1}
Hundredths	centi	c	0.01	10^{-2}
Thousandths	milli	m	0.001	10^{-3}
Millionths	micro	μ	0.000 001	10^{-6}

TABLE 1.2 DERIVED UNITS

Measure	Name of unit	Symbol for unit	Definition of unit
Force	newton	N	$kg.m.s^{-2}$
Energy	joule	J	$kg.m^2.s^{-2}$
Pressure	pascal	Pa	$N.m^{-2}$
Area	square metre	m^2	
Volume	cubic metre	m^3	
Density	mass per volume	ρ	$kg.m^{-3}$
Liquid volume	litre	L	$10^{-3}.m^3$

for m × m and because it is positive it indicates that the unit appears as a numerator in a fraction. Now, what about the $^{-2}$? It is called a negative exponent. It is short for s × s, but, being negative, it indicates that the unit appears in the denominator in a fraction or as $1/s^2$. So $kg.m^2.s^{-2}$ can also be written as $\dfrac{kg.m^2}{s^2}$, and $N.m^{-2}$ as: $\dfrac{N}{m^2}$.

SUPPLEMENT ON UNITS: HUMAN CONVENIENCE

The choice of any system of units is determined by four factors: it must be convenient to use, it must be easily reproducible by all users, it must be based on uniform, regular phenomena and it must establish a scale with uniform segments. The units we use are human inventions. We don't use them because we have to, but because they are convenient and useful. Our use of the metre shows this quite clearly. First of all, a metre is a good length for everyday measurements. Because it's about the length of an adult human leg, or one adult stride, it is convenient to talk about the length of a standard hospital bed as 2 metres, for example. The kilometre is a reasonable unit for common distances involved in travel and the centimetre works well for short lengths down to the limits of human vision.

Second, the whole point of the metre being an international unit is that we can compare lengths from one country with those from another. Each nation must have either its own exact copy of the metre, or be able to make one easily and precisely. In order to do this, the metre is now defined in such a way that any properly equipped laboratory can 'make' a metre. For example, it has been agreed that the modern metre is the distance a beam of light travels in only 1/299 729 458 seconds. This is easily measured in a properly equipped lab and allows accuracy down to incredibly short lengths if required.

Most of the seven basic units are now defined by events that happen within atoms or between atoms. This is because atoms are so exactly alike that what they do is perfectly regular and predictable. In the case of the metre, that measurement provides us with a uniform length which we represent with a metre-rule. (In this sense, a metre-rule is a convenient 'copy' of the distance travelled by that light beam.) It's a scale which we can be sure others are using as well.

The atomic basis for the SI system is constantly being refined, and changes to the way certain phenomena are measured also frequently occur. For example, the units in which radiation exposure is measured have changed from RADs and

REMs to sieverts and grays; a fuller discussion of this is given in Chapter 12, on nuclear radiation.

1.4 SIMPLE METRIC CONVERSIONS

It is often necessary for a nurse to be able to change from one unit to another, more conveniently sized one. Since the SI system is a decimal one, based on units of 10, such conversions are quite simple. For example:

1. Convert 2.4 M units of penicillin to units of penicillin.
   Since M means 'mega', which is the prefix meaning million, 2.4 M units is 2.4 million units, written 2 400 000 units.

2. How many grams are there in 63 mg?
   Since m means 'milli', which is the prefix meaning 1/1000 (or 0.001 or one thousandth), 63 mg means 63/1000 grams, or 0.063 g.

3. The doctor orders a 500 μg dose of a drug. Only 1 mg tablets are on hand. How many would you give?
   In this case we need to know how many 500 μg doses are in each 1 mg tablet. First of all, remember that m is the prefix for one millionth, or 0.000 001; it is 1000 times smaller than a milli. That means that 1 mg is the same as 1000 μg. So, a 1 mg tablet is the same as a 1000 μg tablet. Since the doctor wants a 500 μg dose and 500/1000 = 1/2, you would have to give half a tablet.
   Note: nurses may often see 500 μg written as 500 mcg.

The key to all metric conversion problems is to ask yourself: will the answer be bigger or smaller than the number you start with? When you change from a larger unit, such as kilograms, to a smaller unit, such as grams, the answer must be larger. For example, 2 kg = 2000 g. When you change from a smaller unit to a larger, the answer must be smaller. For example, 25 mL = 0.025 L.

At the end of this chapter are some more examples and problems for self-assessment.

Powers of 10

Some numbers that a nurse encounters are too large, or too small, to be easily handled by the prefixes of the SI system alone, so they are written as powers of 10. The idea behind this is quite simple, though some practice is probably necessary before you become familiar with it. You may recall from your earlier schooling that $2^2 = 2 \times 2$ or that $5^2 = 5 \times 5$ or that $3^3 = 3 \times 3 \times 3$. In each case the ^{2} or ^{3} is called the positive exponent or power. The same concept applies in the following situation:

$$10^3 = 10 \times 10 \times 10 = 1000; \text{ therefore, } 10^3 = 1000$$

Notice that the exponent ^{3} equals the number of zeroes in 1000. It also shows the number of places the decimal point has moved towards the right from its initial position immediately to the right of the 1. In other words, you can say:

10^3 means a 1 followed by 3 zeroes
10^5 means a 1 followed by 5 zeroes
10^{12} means a 1 followed by 12 zeroes = 1 000 000 000 000

Clearly, 10^{12} is a shorter way of writing such a large number. Many times you will see very large numbers, such as the number of red blood cells in the blood, written in such a shorthand form. For example, the average number of red blood cells in every mm^3 of blood could be written as 5×10^6. This is short for $5 \times 1\,000\,000 = 5\,000\,000$.

Small numbers are handled in the same way. In this case, the negative exponent or power tells you the number of places the decimal point has moved towards the left from its original position to the right of the 1: 10^{-3} means 0.001 and 10^{-6} means 0.000 001. The fact that the exponent is a negative number (e.g. $^{-6}$) tells you the number is less than one. In mathematical terms, a negative exponent such as $^{-6}$ refers to a reciprocal. The reciprocal of any number, N, is $1/N$ and $N^{-1} = 1/N$. Recall what we said earlier, after Table 1.2, about negative exponents. Here are two examples of the use of powers of 10 for small numbers:

$$8 \times 10^{-4}\,g \quad = 8 \times 0.0001\,g \quad = 0.0008\,g \; or$$
$$= \frac{8}{10^4}\,g \quad = \frac{8}{10\,000}\,g$$
$$1.8 \times 10^{-7}\,mL \quad = 1.8 \times 0.000\,000\,1\,mL \quad = 0.000\,000\,18\,mL \; or$$
$$= \frac{1.8}{10^7}\,mL \quad = \frac{1.8}{10\,000\,000}\,mL$$

Notice that for negative exponents, the larger the exponent the smaller the number; that is, 10^{-7} is much smaller than 10^{-4}.

There are more of these examples and some problems at the end of this chapter. Table 1.3 gives you some essential metric information.

TABLE 1.3 METRIC RELATIONS

1 L of water has a mass very close to 1 kg and a volume of 1000 cm³
1 mL of water has a mass very close to 1 g and a volume of 1 cm³
1 tonne (t) is defined as 1000 kg

Worked examples of metric calculations

Length

1. Convert the following to metres: (a) 349 cm (b) 4 mm (c) 4256 µm
   (a) 349 cm = 349/100 = 3.49 m
   (b) 4 mm = 4/1000 = 0.004 m
   (c) 4256 µm = 4256/1 000 000 = 0.004 256 m

2. Convert the following to millimetres: (a) 9.5 m (b) 215 µm (c) 11.52 cm
   (a) 9.5 m = 9.5 × 1000 = 9500 mm
   (b) 215 µm = 215/1000 = 0.215 mm (because 1 mM = 1000 µm)
   (c) 11.52 cm = 11.52 × 10 = 115.2 mm

Volume

3. Convert the following to mL: (a) 6.35 L (b) 3.8 µL
   (a) 6.35 L = 6.35 × 1000 = 6350 mL
   (b) 3.8 µL = 3.8/1000 = 0.0038 mL

Mass

4. Convert the following to grams: (a) 7.54 mg (b) 21.5 kg (c) 0.15 µg
   (a) 7.54 mg = 7.54/1000 = 0.007 54 g
   (b) 21.5 kg = 21.5 × 1000 = 21 500 g
   (c) 0.15 µg = 0.15/1 000 000 = 0.000 000 15 g

Metric problems with specific nursing applications

1. Convert 1.5 megaunits of penicillin to units of penicillin.
   Answer: 1.5 MU = 1.5×10^6 = $1.5 \times 1\ 000\ 000$ = 1 500 000 units.

2. Prescribed dosage is 500 µg. On hand are 1 mg tablets. How many do you give?
   Answer: 500 µg = 500/1000 = 0.5 mg; therefore, 1/2 tablet.

3. Prescribed dosage is 250 µL per hour. Supplied is a 100 mL bottle. How long will it last?
   Answer: 100 mL = 100×1000 = 100 000 µL
   100 000/250 = 400; therefore, it will last 400 hours.

4. How many pascals is a diastolic blood pressure of 15.7 kPa?
   Answer: 15.7 kPa = 15.7×1000 = 15 700 Pa.

Questions

Level 1

1. Identify each of the following as standard or derived units: m^3, kg/m^3, s, kg, cm/s.
2. Arrange in order of increasing size: centimetre, kilometre, metre, millimetre.
3. What is the mass of 5000 cm^3 of water?
4. What are the two parts of every measurement?
5. Skin has been measured to be 0.000 08 m thick. Express this number in powers of 10.

Level 2

6. State some of the advantages of the SI system of measurement.
7. Try to find out the normal range of body temperatures.
8. Convert 24 m to: (a) cm (b) km.

Level 3

9. A famous scientist, Lord Kelvin, once said that he could only truly understand something if he could measure it. That is, he had to be able to assign a number and a unit to, say, length, before he felt he truly knew what length means. Do you agree? What sort of things do you feel do not require some type of measurement to understand?

SUPPLEMENTARY DRUG CALCULATION PROBLEMS

1. The client has been prescribed 10 mg of Valium. On hand you have only 0.005 g tablets. How many will you give?
2. The prescription is for 1 mg of Resperine. The label on the vial reads 2.5 mg per mL. How many mL should be given?
3. (a) Nalidixic acid 4000 mg is prescribed, to be administered in four equally divided doses. How many mg are given in one dose?
   (b) The tablets in the pharmacy supply are 500 mg/tablet. How many tablets per dose?
4. Atropine sulphate 600 μg is prescribed. Pharmacy stock is 1.2 mg/mL. How many mL are to be given?

ANSWERS TO SUPPLEMENTARY DRUG CALCULATION PROBLEMS

1. Step 1: change 10 mg to g

    10 mg = 10/1000g = 0.01 g.

    Step 2: divide amount needed by amount available

    0.01 g/0.005 g = 2; therefore, 2 tablets are needed.

2. Note here that 2.5 mg per mL means that 2.5 mg of the drug is dissolved in every 1 mL of the solution.

    Divide amount needed by amount available:

    1 mg ÷ 2.5 mg/mL = 0.4 mL to be given.

3. (a) 4000 mg/4 = 1000 mg per dose.

    (b) 1000 mg/500 mg per tablet = 2 tablets to be given.

4. Step 1: change 600 μg to mg

    600 μg = 600/1000 mg = 0.6 mg.

    Step 2: divide amount needed by amount available

    0.6 mg/1.2 mg per mL = 0.5 mL required.

2

ORGANISATION OF THE HUMAN BODY

STRUCTURE AND FUNCTION, HOMEOSTASIS AND TEMPERATURE CONTROL

Chapter outline

2.1 **Structure and function.** Your objective: to understand the meaning of these terms and how they help us understand how the body works in health and illness.

2.2 **A typical human cell.** Your objective: to understand basic cell structure and the various types of cells in the human body.

2.3 **Tissues and organs.** Your objective: to be able to describe the defining characteristics of tissues and organs and list the principle ones found in the body.

2.4 **Systems of the body.** Your objective: to be able to define and list the major systems in the human body.

2.5 **Homeostasis.** Your objective: to be able to define the term precisely and apply it correctly in terms of how the body maintains control over its structures and functions.

2.6 **Heat energy and homeostasis.** Your objective: to understand the ways in which the body, using homeostasis, controls its temperature.

INTRODUCTION

This chapter contains necessary background information on the overall organisation of the human body from a biological point of view. It introduces the idea of homeostasis, which refers to the tendency of our bodies to maintain a constant internal environment. It also examines the control of heat in the body as an example of homeostatic control.

Case Study: Mrs Carmel H.

Mrs H. has been admitted to hospital with pneumonia and one of her major symptoms is fever. The daily temperature readings, which were fluctuating between 36°C and 40°C, have started to settle down to 37°C. When feverish, she is treated with sponge baths of cool water and with aspirin, an example of an antipyretic drug ('antipyretic' means 'anti-fever'). Her fluid intake has been increased to counter the effects of sweating. Of course, the pneumonia infection is also being treated with antibiotics, as it is the source of the symptoms of pneumonia, including the fever.

Clinical notes for Mrs H.: Common nursing goals are to improve oxygenation, maintain fluid balance, prevent infection, and to provide adequate nutritional intake. These are all directed at helping Mrs H. maintain homeostasis. Fluid balance will be assessed measuring urinary output, with an expected outcome of greater than 30 mL/hr and a specific gravity of between 1.010 and 1.030. IV fluid therapy may be needed.

2.1 STRUCTURE AND FUNCTION

These are crucial concepts for understanding how the body works and fortunately they are quite simple. Function simply refers to what any part does. Thinking of it like that saves a lot of confusion. For example, many people would say the function of the heart is to pump blood around the body. It's more correct to say that the function of the heart is to beat; to contract rhythmically over and over for as long as possible. If by doing that, blood is pumped around the body, that's fine. But the heart's function, what it does, is only to beat. Similarly, the function of the eye is to convert light energy to electrical energy in the optic nerves; the function of a muscle is to contract when stimulated.

The structure of the heart allows it to perform its function. It has all the right parts and connections between parts, to allow it to beat. For example, it has the appropriate type of muscle, an energy supply and the necessary information about when to beat. The connection between structure and function is clearly very close. As an everyday example, screwdrivers come in two shapes, in order to match structure to function. Similarly, for the heart to function properly, it needs the right structure. Many conditions of ill-health occur because the cells, tissues, organs or systems shortly to be described no longer perform their functions, in many cases because of damage to their structures.

The organisation of the human body

To understand how Mrs H., or any of us, can control body temperature without medication, it is first necessary to have some idea of overall body organisation. Over and over again, throughout this book, we will see how the body as a whole behaves as it does because of what is happening to its smaller parts, the **cells**. When we treat the whole person, much of what we do is aimed at influencing those tiny structures that compose the whole.

The human body is not just a jumble of parts. It is composed of a large number of specialised components, which work together in a co-ordinated way. There are several levels of organisation in the body, from the simplest to the most complex. This is referred to as a hierarchical organisation and it is shown in Figure 2.1.

Cell
The body's smallest independent unit of life. Cells have complex structures that allow them to carry out their diverse functions.

2.2 A TYPICAL HUMAN CELL

Health and illness are words that usually describe the condition of the entire person. From the physiological point of view, it can be said that the health of the

FIGURE 2.1 Levels of organisation of the human body

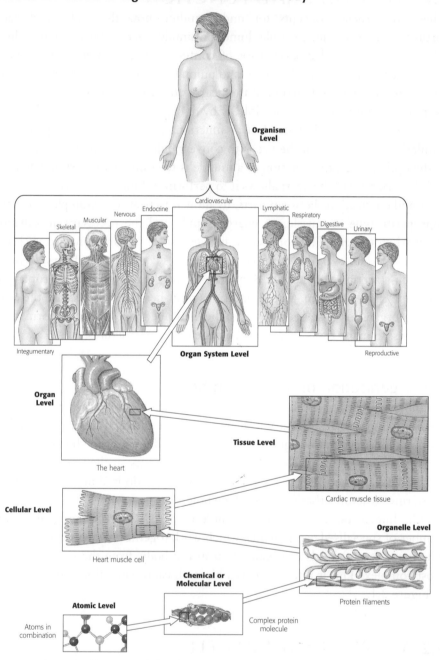

Organism Level

Cardiovascular

Skeletal
Muscular
Nervous
Endocrine
Lymphatic
Respiratory
Digestive
Urinary

Integumentary

Organ System Level

Reproductive

Organ Level

The heart

Tissue Level

Cardiac muscle tissue

Cellular Level

Heart muscle cell

Organelle Level

Protein filaments

Chemical or Molecular Level

Atomic Level

Atoms in combination

Complex protein molecule

Source: *Fundamentals of Anatomy and Physiology*, 5th Ed, Martini et al, 2001, p. 5. © Reprinted by permission of Pearson Education Inc., Upper Saddle River, N.J.

person depends on the health of his or her parts. For humans, the smallest living parts are the cells, which range in size from about 10 to 30 micrometers (10 μm–30 μm) in diameter. And, speaking simplistically, when they're well, we're well. An adult contains roughly 80×10^{12} cells, most of which are alive (though some, like the upper layer of the skin and hair, are dead). It is their individual lives and their co-operative effort that make up what we call living. Figure 2.2 shows the main structures found in a typical human cell. Many of these will be discussed in greater detail in later sections of this book.

Each structure in the cell has a specific function to perform. The membranes, for example, do not just hold the cell or nucleus together. They are capable, to some extent, of controlling what substances enter and leave the cell. They can also respond quickly to chemical messages, in the form of substances called hormones, which are sent via the bloodstream throughout the body from

FIGURE 2.2 The main structures of a typical cell

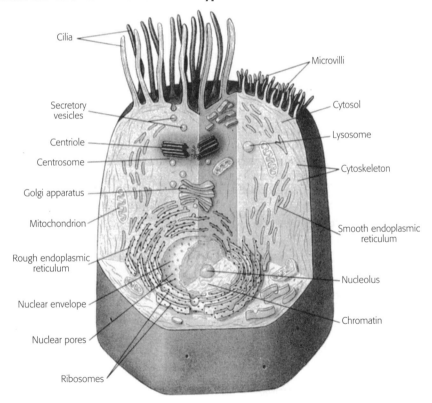

Source: *Fundamentals of Anatomy and Physiology*, 2nd Ed, Martini, F., 1992, p. 67. © Reprinted by permission of Pearson Education Inc., Upper Saddle River, N.J.

such places as the adrenal glands above the kidneys. These messages may cause the cell membrane to allow new substances, or different amounts of familiar substances, to enter or leave the cell; that is, they modify its structure and therefore influence its function.

If there is a sudden change in the outside environment, this ability of the cells to respond rapidly is clearly important. For example, physical danger, such as the threat of an automobile accident, makes the adrenal glands release hormones called adrenaline and noradrenaline, which produce immediate changes in body activity (in some American textbooks these hormones are referred to as epinephrine and norepinephrine). The heart beats faster and the supply of blood to the muscles is increased, among other things. Changes to the internal environment, such as those brought about by invading bacteria, can also prompt such cellular responses. This illustrates how the whole body is linked to the ability of its cells to respond to changes in their environment. In Mrs H.'s case, her ability to control her fever is linked to just such changes in her cell membranes, as we will see.

Keep in mind that there are many different types of cells in the body—for example, liver cells, heart muscle cells and nerve cells. Each type has a particular function, structure and physical appearance. When we look at particular parts of the body in more detail later, we'll examine more closely how each cell type is unique.

2.3 TISSUES AND ORGANS

Tissue
A collection of cells of the same type, performing the same function (e.g. muscle tissue).

A **tissue** is a collection of cells of the same type, specialised to perform a particular function or group of functions. For example, muscle tissue is composed of muscle cells, all of which have similar structure and function. It is important to learn the technical names for the body's main tissue types, set out in Figure 2.3.

FIGURE 2.3 Main tissues of the body

EPITHELIA	CONNECTIVE TISSUES	MUSCLE TISSUE	NEURAL TISSUE
– Cover exposed surfaces – Line internal passageways and chambers	– Fill internal spaces – Provide structural support – Store energy	– Contracts to produce active movement	– Conducts electrical impulses – Carries information

Source: *Fundamentals of Anatomy and Physiology*, 2nd Ed, Martini, F., 1992, p. 124. © Reprinted by permission of Pearson Education Inc., Upper Saddle River, N.J.

The word 'organ' comes from the Greek *organon*, which means 'instrument'. **Organs** in the body are, in a sense, instruments for performing a specific task. The heart, for example, is an organ whose function is to beat, as we've said, though in relation to the body as a whole we might say its task is to pump the blood. The technical definition of an organ is: a collection of two or more different types of tissues, which work together to perform a particular function. There are many organs in the body: heart, lungs, stomach and bones are common examples. The stomach is considered an organ because it contains many different types of tissues—nervous, epithelial, muscular, and others—all working together for the purpose of performing part of the task of digestion. In the case study that introduced this chapter, Mrs H. has pneumonia, simply defined as an inflammation of the lungs. The lungs are an organ, comprised of epithelial and connective tissues, as well as others.

Organ
A combination of two or more tissues into one structure, carrying out a single function (e.g. the heart).

2.4 SYSTEMS OF THE BODY

A **system** in the body consists of a group of organs working together to perform specific tasks. These organs do not have to be physically connected together, as are the tissues in an organ. They can be far apart in the body. For example, both the heart and the smallest capillary in the little toe are part of the circulatory system. They share the same tasks, one of which is circulating the blood around the body. Some organs are part of more than one system and so it is necessary to be flexible in your thinking about the role of any organ in the body. Figure 2.4 lists the major body systems and their main tasks, but the organs of each system have not been included.

Mrs H.'s lungs are part of the respiratory system. One of the respiratory system's major tasks is to supply the body with oxygen and dispose of carbon dioxide. Other organs that are part of this system are the nose, pharynx, larynx, trachaea and bronchi.

System
A combination of organs carrying out a major task, such as the circulatory system. Organs may be part of more than one system.

2.5 HOMEOSTASIS

Cells can be thought of as the ultimate traditionalists; they hate change. They like being where they are (e.g. deep in the liver) because of the favourable conditions there (their environment) and in fact can't survive for long if that environment changes even slightly. They need protection from all types of variations (e.g. in such things as temperature, types of nutrients and amount of oxygen) that occur in their environment. In other words, the body needs to control the environment in which its cells live.

FIGURE 2.4 Major systems of the body

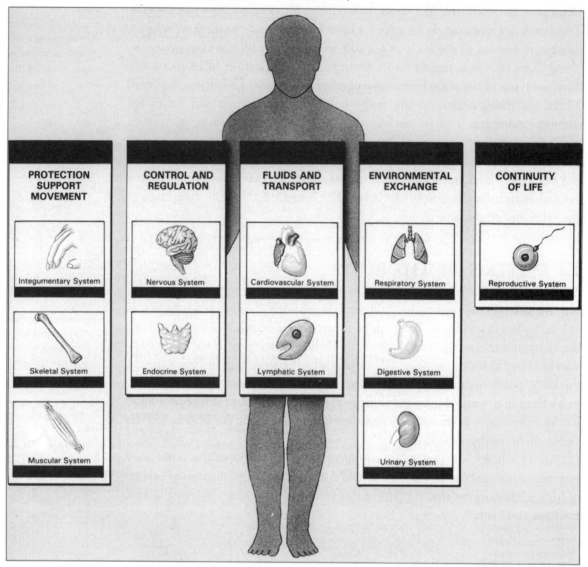

Source: *Fundamentals of Anatomy and Physiology*, 2nd Ed, Martini, F., 1992, p. 168. © Reprinted by permission of Pearson Education Inc., Upper Saddle River, N.J.

The technical term for the tendency to control the internal environment is **homeostasis**, which literally means 'same state'. Life is possible only when conditions inside the cell (or body) remain relatively steady, an idea first pointed out over a century ago by the French physiologist Claude Bernard. In his words,

'All the vital mechanisms, varied as they are, have only one object, that of preserving constant the conditions of life in the internal environment'.

The outside environment may or may not be hospitable to the cell or organism living in it. One way to think of it is as a war between inside and out and, as long as inside wins, we live. The body uses a great many very complicated and precise mechanisms to win this war. When they fail to work properly, the health of the person declines. When, as a nurse, you measure particular body conditions, such as the client's temperature, you are getting information about the internal environment. When a measurement falls outside the normal range of values, it means homeostasis is not being maintained and the person's health is affected in some way.

Because control is not precise, the internal environment is only relatively constant. This means there is a small range of values that are normal for that measurement. In Chapter 1 the idea of normal values was discussed. Table 2.1 gives the normal range of values for some of these measurements.

The way this control is achieved is by co-operation—the different parts of the body working together. A cell deep in the liver can't do its job without being supplied with food, water, energy and raw materials from elsewhere. And, in turn, the suppliers of that cell need the products produced by it for their survival. By working together, they help to maintain an optimum environment. But how do they achieve such co-operation?

In general, homeostasis works as follows. A **sensor**, somewhere within the body or on the surface, has the task of detecting a possibly dangerous change in its surroundings. Maybe the temperature of the blood has moved outside the normal range, or the amount of oxygen in the blood is lower than normal. The sensor sends information about this change to another organ, called the integrating centre, which first interprets the information and then directs some other part of the body, called an **effector**, to make some change in its normal functioning.

Homeostasis
The term used to describe the complex and numerous activities carried out by the body to maintain a relatively constant internal environment (e.g. relatively constant temperature).

Sensor
A structure in or on the body which responds to a change in its environment by signalling to other organs or systems (e.g. the hypothalamus and temperature, the eye and light).

Effector
Structures in the body which respond to signals from a sensor or an integrating centre like the brain, by altering their normal function (e.g. muscles shivering).

TABLE 2.1 SELECTED NORMAL VALUES OF BLOOD

Substance	Normal value
Blood volume	8.5–9% of body weight in kg
Calcium	8.5–10.5 mg/mL
Iron	50–150 µg/100 mL
Red blood cell count	4.8–5.4 million/mm^3

Negative feedback loop
A common homeostatic
control mechanism. A
stimulus, or change away
from normal in the
internal environment
(e.g. a decrease in blood
temperature) leads to a
particular action being
taken by the body (e.g.
shivering). This action by
the effectors makes the
detected change decrease
(e.g. the blood
temperature rises towards
normal). This decrease in
the change causes a
reduction in the body's
reaction to the change
(e.g. shivering slows
down, then stops).

This alteration of normal function usually tends to reverse whatever changes in the sensor's surroundings started the whole process. For example, effector muscles might be told to start shivering to generate heat, or the rate of breathing might be increased. If these actions succeed in reversing the change, the sensor should start noticing that things are back to normal in its neighbourhood. It reduces or stops its message to the integrating centre, where that message is interpreted as a decrease in change. The effectors are then told they can go back to normal operations.

Let's look at an example of homeostasis that relates to Mrs H.'s condition of fever. Figure 2.5 presents a specific example of what is technically referred to as a **negative feedback loop**.

The action the body takes leads to a decrease in the possibly dangerous

FIGURE 2.5 A negative feedback loop and the control of body temperature.
Body temperature is regulated by a control centre in the brain that functions like a thermostat. It normally accepts a temperature range of 36°C–38°C. If the temperature falls below 36°C, heat is conserved by restricting blood flow to the skin, and more heat is generated through shivering. If the temperature climbs above 38°C, heat loss is increased through enhanced blood flow to the skin and sweating. In each instance a variation outside normal limits triggers an automatic feedback response that corrects the situation. This mechanism is called negative feedback.

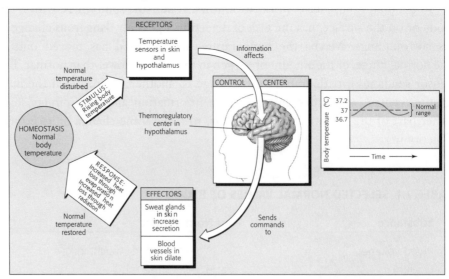

Source: *Fundamentals of Anatomy and Physiology*, 5th Ed, Martini et al, 2001, p. 13. © Reprinted by permission of Pearson Education Inc., Upper Saddle River, N.J.

changes the sensor first detected. Other examples of this will be discussed later, but the important points about homeostasis are as follows:

1. Cells can survive only under certain conditions such as correct temperature, level of oxygen, and the presence of nutrients.

2. Homeostasis is the name given to the complex and numerous activities the body uses to maintain these conditions in the body.

3. An important aspect of homeostasis is the co-operation between all parts of the body and this is most commonly achieved by negative feedback loops:
   □ First, a sensor detects a change in the internal environment.
   □ An integrating centre directs the body to respond in some way.
   □ The appropriate organ or system follows those directions.
   □ The sensor detects that the change has decreased.
   □ The organs or systems return to normal functioning.

4. Small adjustments occur all the time to maintain health, as we interact with the outside environment and as the internal environment fluctuates due to its own activities.

The integrating centres most responsible for giving directions to other parts in homeostasis are the brain and the endocrine glands. Another example of homeostasis in action is the negative feedback control of insulin secretion, as shown in Figure 2.6.

The sensors are detecting changes in the amount of a type of sugar called glucose in the blood, which is the environmental factor being controlled.

SUPPLEMENT ON HOMEOSTASIS

Although negative feedback is perhaps the most common way homeostasis is maintained, it's important to remember that it is only one part of a very complicated, subtle process of control. Looking at it in greater detail, we can see two other parts of the picture that help us to understand how the body regulates its internal environment.

1. *Antagonistic effectors*. Antagonists are opponents that try to reverse whatever the other has done; for example, normal body temperature is about 37°C,

FIGURE 2.6 The regulation of blood glucose concentration

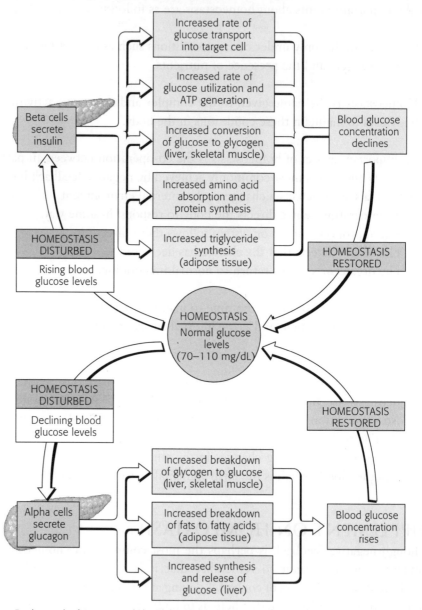

Beta cells secrete insulin

Increased rate of glucose transport into target cell

Increased rate of glucose utilization and ATP generation

Increased conversion of glucose to glycogen (liver, skeletal muscle)

Increased amino acid absorption and protein synthesis

Increased triglyceride synthesis (adipose tissue)

Blood glucose concentration declines

HOMEOSTASIS DISTURBED
Rising blood glucose levels

HOMEOSTASIS RESTORED

HOMEOSTASIS
Normal glucose levels
(70–110 mg/dL)

HOMEOSTASIS DISTURBED
Declining blood glucose levels

HOMEOSTASIS RESTORED

Alpha cells secrete glucagon

Increased breakdown of glycogen to glucose (liver, skeletal muscle)

Increased breakdown of fats to fatty acids (adipose tissue)

Increased synthesis and release of glucose (liver)

Blood glucose concentration rises

Source: *Fundamentals of Anatomy and Physiology*, 5th Ed, Martini et al, 2001, p. 606. © Reprinted by permission of Pearson Education Inc., Upper Saddle River, N.J.

though it varies slightly around the body. If it falls below this value, one effector, the muscles, tries to raise body temperature by generating heat through shivering. As body temperature rises to normal, homeostatic control does not switch off the shivering. Instead, an antagonistic or opposing effector, sweating, switches on and helps prevent the body overheating. The combination of these two effectors, working to produce opposite effects, fine-tunes the body's temperature level. Other effectors are also involved in the control of body temperature (e.g. vasodilation and vasoconstriction, discussed later in this chapter), all of which contribute to maintaining a relatively constant 37°C.

2. *Positive feedback loops*. In this case, the effectors cause a further increase in the change that stimulated the sensors. It is as though, as the body grows warmer, the amount of shivering increases rather than stops. Not surprisingly, such positive feedback loops are not common, as they remove the body further from homeostasis. They are, however, often used as one part of a larger negative feedback loop; for example, the clotting of blood. If blood clots were too easy to form, there would be a constant risk of them forming inside our circulatory system, so a series of many steps is involved in reducing this risk. One step is not to be taken until the previous step has occurred. This series of steps is a positive feedback loop: each step does not reverse the change but instead enhances it, and each step activates another until the whole series is complete and the clot forms. Of course, the clot itself is only one factor in a larger negative feedback loop which is responsible for controlling the body's fluid volume. Figure 2.7 shows the blood clotting positive feedback loop.

FIGURE 2.7 Blood clotting positive feedback loop.
In positive feedback, a stimulus produces a response that reinforces the original stimulus. Positive feedback is important in accelerating processes that must proceed rapidly to completion. In this example, positive feedback enhances the clotting process, which seals breaks in any blood vessel walls and prevents blood loss.

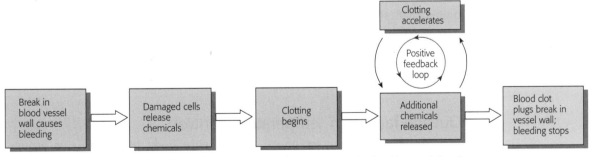

Source: *Fundamentals of Anatomy and Physiology*, 5th Ed, Martini et al, 2001, p. 15. © Reprinted by permission of Pearson Education Inc., Upper Saddle River, N.J.

Some technical terms associated with homeostasis

- *Stimulus*—the change in state detected by the sensors.
- *Response*—the action taken by the effectors to return the body state to normal.
- *Effector*—the organ or system that acts to restore homeostasis.
- *Integrating centre*—a term often used for the organ that determines the range of values that are to be maintained, analyses the messages from the sensors and then sends the appropriate command to the effectors.
- *Dynamic equilibrium*—'dynamic' means here that homeostasis is maintained by the body actually working at it; 'equilibrium' means that homeostasis is a balance between changes that tend to either increase or decrease the body's preferred, normal values.
- *Autonomic*—without our conscious control; most homeostatic mechanisms work without us having to think about it.
- *Stress*—sometimes used to stand for a condition of the body when it is not in homeostasis. It is caused by stressors—all those factors that act on the body to change its steady state, such as accidents, infection, poor diet, psychological distress, and many more.

2.6 HEAT ENERGY AND HOMEOSTASIS

Perhaps now we can understand Mrs H.'s situation more fully. Her bacterial infection is causing her fever. The bacteria are producing chemical toxins which upset the internal 'thermostat', or temperature control, located in an organ called the hypothalamus, to be discussed shortly. In effect, her temperature level has been set too high, at, say, 39°C. By fighting the bacteria with antibiotics, we are attempting to remove the cause of that fever. But we also treat her symptoms by cool sponging and the administration of antipyretics. And finally, her body is attempting to maintain a constant internal temperature by sweating and shivering. She is shivering, despite her fever, because, with her thermostat set higher than normal, her body acts as if the new temperature level is the correct one. Let's have a closer look at the last two ideas mentioned above: our external treatment of her symptoms and her internal homeostatic control of temperature.

Outside factors in temperature control

For Mrs H. and the rest of us, normal core body temperature is around 36–37°C. When healthy, we can maintain that temperature despite changes in the environment around us. The key idea here is that any heat gained by the body, either from the outside (such as the sun), or from within (by, for example, shivering), must be equalled by a heat loss from the body. Similarly, any heat lost by the body to the outside (as when we swim in cold water) must be equalled by a gain in heat by the body. In short:

Heat loss = Heat gain

Heat can be exchanged between the body and the outside environment in four ways: radiation, conduction, convection and evaporation.

1. Radiation

Loss or gain by radiation means that the heat is in the form of infrared rays, often called heat rays. These are a form of electromagnetic radiation (see Chapter 9) invisible to the eye, but detectable by the skin as heat. All objects radiate infrared rays. If our body is warmer than the surrounding room, it will radiate heat into the room, thus suffering a heat loss. In fact, 60% of human heat loss is by radiation, though clothing affects this. Obviously, the body can gain heat by radiation if it is in a room that is warmer than it is, or close to a source of radiant heat such as an infrared or heat lamp.

2. Conduction

This is the transfer of heat from warmer to cooler objects by direct touching, as when you sit on a cold chair. Usually only small amounts of heat are gained or lost this way. However, it can be important, as we saw with Mrs H. The cool water on her skin removed some of the heat from her body by conduction; Mrs H. got cooler, the water got warmer.

3. Convection

If we had left the cool water on Mrs H., both her skin and the water would soon reach the same temperature. As the skin transferred heat to the water, the water temperature would rise until it matched the temperature of her skin. The usefulness of the water in cooling by conduction would be limited. However, by removing the warmed water and replacing it with new, cool water, the process of cooling can continue. If the source of heating or cooling is continually replaced, we have convection. The source is usually air, as in a fan, or water, and since both are moving fluids we speak of convection currents. Even in relatively still air, we can lose 12% of our heat by convection.

4. Evaporation

This works only to remove heat from the body. As you may know, it takes heat to evaporate water; that is, change it from liquid to gas. When the water in sweat evaporates from the skin, it takes away from the body about 2.4 kJ (or 0.58 kilocalories) of heat for each gram of water used. Even when we're not visibly sweating, we lose water from the skin and lungs at a rate of about 600 mL per day. By controlling the rate of sweating, we can regulate the amount of heat lost this way. This is why Mrs H. is perspiring; she is trying to lose excess heat produced by the fever. Because she is ill and her thermostat is too high, she is also trying to produce heat by shivering, clearly two things we don't normally do at the same time. The sponging helps too: it takes more heat to evaporate the cold water from her skin than it does to evaporate the warmer sweat.

Figure 2.8 shows all these factors at work.

FIGURE 2.8 Heat gain and loss in temperature control

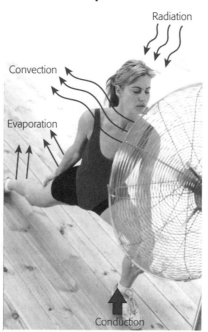

Source: *Fundamentals of Anatomy and Physiology*, 5th Ed, Martini, et al, 2001, p. 934. © Reprinted by permission of Pearson Education Inc., Upper Saddle River, N.J.

Of course, they are all affected by our behaviour, too, especially the wearing of appropriate clothing. Clothing holds a layer of air near the body: this layer is quickly warmed to the same temperature as the skin and prevents convection. Looser clothing allows convection currents to move such warmed air away from the skin, thus cooling it.

Internal control of temperature

Earlier, when discussing negative feedback loops, we mentioned sensors for detecting change in the internal environment. For temperature, this sensor occurs in that part of the brain called the hypothalamus, the thermostat mentioned earlier, set normally at around 37°C. This organ contains a large number of heat-sensitive neurones, or nerve cells, which become more active as the temperature of the blood passing by them increases above the normal setting. (A smaller number of cold-sensitive nerves are also present there, to detect blood temperatures below the normal setting.) The increased activity of these high-temperature neurones brings about several changes to different effectors around the body.

Vasodilation

This means that the blood vessels in the skin become dilated, or enlarged. More blood is then close to the surface of the body, where it can lose its heat through radiation, conduction or convection. This is visible to the eye as 'flushing' of the skin. Vasodilation can increase heat loss by up to eight times. It is interesting that it occurs because the increased activity of the hypothalamus turns off vasomotor tone, which is responsible for a slight degree of narrowing of the blood vessels, the normal condition of the arteries and veins in the skin. The opposite of vasodilation is vasoconstriction, which is a reaction to cold. The diameter of the blood vessels decreases, keeping the blood away from the surface of the body. It is therefore a greater constriction than the normal condition of vasomotor tone.

Sweating

The sweat glands in the skin are connected to the brain by the autonomic nervous system. A detailed discussion of the autonomic system is outside the scope of this book, but essentially it controls the automatic, involuntary acts of the body, such as the rate of heartbeat, the contractions of the stomach and blood vessel diameter. When the sweat glands are stimulated by the neurones of the autonomic nervous system that originate in the hypothalamus, they secrete sweat.

Piloerection

Piloerection means 'hairs standing on end'. When we get 'goosebumps', we can see our skin reacting to the nerve signals from the hypothalamus and trying to erect our hairs to trap a thicker layer of air next to the skin. Fortunately, clothing does the job much better, but piloerection is still an important mechanism in furred animals.

Decrease or increase in heat production

Activities that generate heat, such as shivering, can be increased or cut down as appropriate. These are sometimes due to changes in body metabolism, mentioned below and discussed more fully in Chapter 4.

Changes in metabolism

The main source of heat within the body is metabolic heat, which is given off as waste by all the chemical reactions taking place within our cells. It's enough to keep our temperature close to the normal 37°C, with some left over. To remain active, an average adult uses about 8.4×10^6 joules (2000 kilocalories) of energy a day. If needed, say during exercise or when shivering from cold, this can be increased up to 3.5 times. Metabolic rate, like temperature, is under homeostatic control. If normal temperature regulation is not maintained by homeostasis, problems obviously arise, as shown by Figure 2.9.

Control of body temperature is another example of homeostasis affecting the whole body by changing what cells do: if blood temperature falls, the hypothalamus can increase the production of a chemical called thyrotrophin-releasing hormone (TRH). This hormone travels to the pituitary gland at the base of the brain. The pituitary is responsible for producing a range of hormones involving growth and homeostatic control. In this case, it produces a thyroid-stimulating hormone (TSH), or thyrotropin. This in turns travels to the thyroid, where it causes the thyroid to produce thyroxine, a hormone which increases the body's metabolic rate. Since this whole process takes up to a week, it is no good for immediate action against excess cooling, but it does show how homeostasis can deal with both short-term and long-term problems.

FIGURE 2.9 Normal and abnormal variations in body temperature

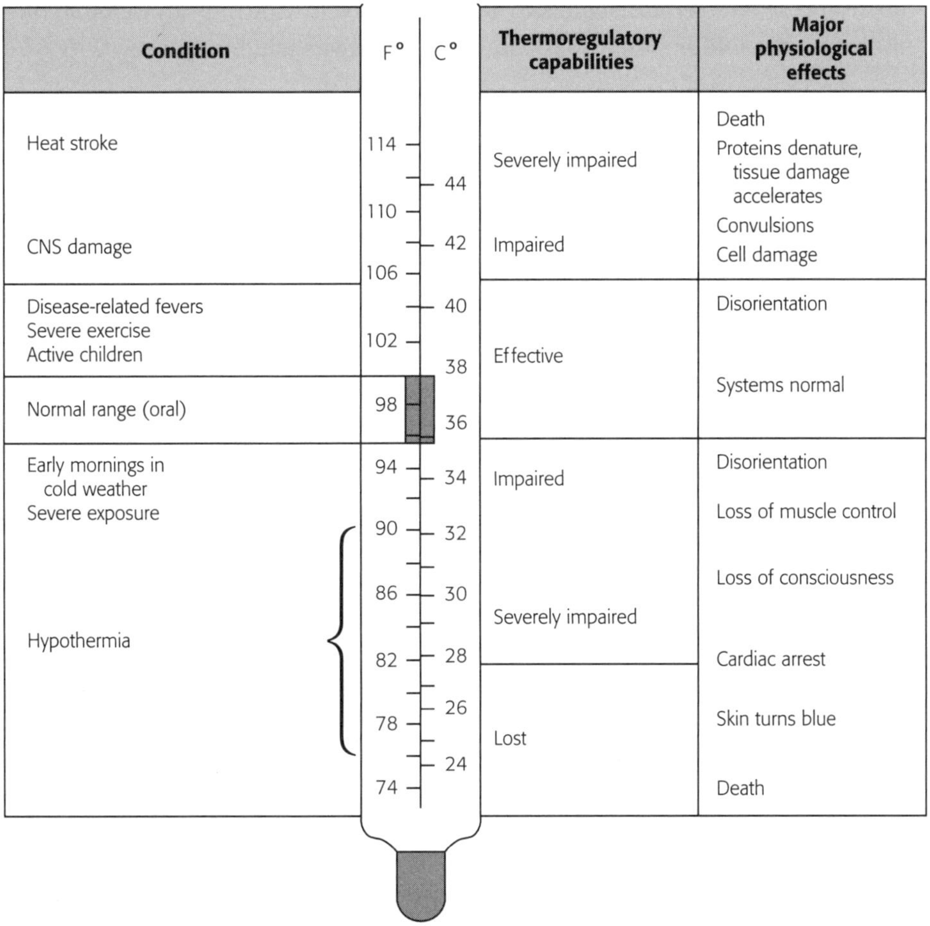

Condition	F°	C°	Thermoregulatory capabilities	Major physiological effects
Heat stroke	114	44	Severely impaired	Death Proteins denature, tissue damage accelerates
CNS damage	110 106	42	Impaired	Convulsions Cell damage
Disease-related fevers Severe exercise Active children	102	40 38	Effective	Disorientation
Normal range (oral)	98	36		Systems normal
Early mornings in cold weather Severe exposure	94 90	34 32	Impaired	Disorientation Loss of muscle control
	86	30	Severely impaired	Loss of consciousness
Hypothermia	82	28		Cardiac arrest
	78	26 24	Lost	Skin turns blue
	74			Death

Source: *Fundamentals of Anatomy and Physiology*, 2nd Ed, Martini, F., 1992, p. 852. © Reprinted by permission of Pearson Education Inc., Upper Saddle River, N.J.

What about those antipyretics that Mrs H. has been given? These contain chemicals that can directly, but only for a short time, influence the sensor cells in the hypothalamus. In effect, they tell the thermostat cells to reset the temperature setting to the normal range. We said earlier that a fever results when the thermostat in the hypothalamus is set too high. To be more precise, this is the result of the presence of chemicals (in Mrs H.'s case, chemicals produced by the bacteria causing her pneumonia) which stimulate the production of prostaglandins in the hypothalamus. Prostaglandins are a group of chemically similar substances which have many functions, one of which is to help regulate body temperature. If they are produced in too great a quantity, they can reset the hypothalamus to too high a level. Aspirin is an example of a substance that inhibits the production of these prostaglandins.

The measurement of temperature

The instrument we most often use for the measurement of body temperature is the clinical thermometer. It is different in two ways from the thermometers you may have used before. First, it is designed to measure temperatures of only a limited range, usually 35–45°C. Second, the tube that contains the mercury or alcohol has a constriction that holds the liquid in position once a measurement is made. Otherwise, the liquid level would start to fall soon after taking the temperature. The nurse therefore has more time to record the measurement, but must then shake the liquid back down for the next reading.

The Celsius scale is the one used in nursing, though you may recall that the Kelvin scale is the SI unit of temperature. In the Celsius scale, the freezing point of water is called 0° and the boiling point of water is called 100°. When placed in freezing water, the level indicated by the mercury in a thermometer is marked with a line and labelled 0. When it is then placed in boiling water, the mercury expands and rises in the tube. When it comes to rest, that level is marked as 100. The distance between the two marks is then divided into 100 intervals of equal size and each of these represents 1 degree.

Questions

Level 1

1. Heat loss = Heat gain. How does this principle apply to the case of Mrs H.? What are her sources of heat gain? Of heat loss?
2. Draw a diagram of the negative feedback loop for a case where a client is suffering from hypothermia and is wrapped in warm blankets.
3. The liver is an organ. How is it distinguished from a cell or tissue?
4. What is meant by the term 'the internal environment'?
5. Define the following terms: receptor, effector, stimulus and response.

Level 2

6. Why do you think body temperature varies slightly over a 24-hour period? When is it lowest? Highest?
7. During exercise, the rate of sweating increases. Is this an example of positive or negative feedback? Explain your choice.
8. Clients with fever are often asked to remain quietly in bed. Why?

Level 3

9. Carefully summarise the care Mrs H. is being given and state why it is effective.
10. State the connection between the failure of homeostasis and ill-health.
11. Why are ice packs sometimes applied to sports injuries such as wrist sprains? Would an infrared lamp be better?
12. The organisation of the body is said to be hierarchical. What does this mean? Use this idea to explain the relationship between the circulatory system and a cell receiving nutrients.

CALCULATION PROBLEM RELATED TO THE CASE STUDY

800 mL of sodium chloride 0.9% (normal saline) is dripping into Mrs H.'s arm at 25 drops/min. The IV set delivers 20 drops/mL. How long will it take her to receive the infusion?

Answer:
800 mL/25 drops/min × 20 drops/mL = 640 min = 10 hr 40 min.

3

THE ORTHOPAEDICS WARD AND THE MUSCULAR/SKELETAL SYSTEM

FORCE, MOVEMENT AND ENERGY IN THE HUMAN BODY

Chapter outline

INTRODUCTION

Some important ideas about force and energy are introduced in this chapter. They are mentioned so early in this text because they relate both to your clients' comfort and to the prevention of injury to you as a nurse. They will also be referred to in later discussion of other body systems and techniques for client care. More specifically, this chapter deals with the concepts of force, pressure, energy, levers and human movement.

Case Study: Mr David M.

As you learn the special skills involved in the activities of daily living on the orthopaedics ward, you are made responsible for the care of David M. He was involved in a car accident two weeks ago and since then has been in balanced skeletal traction with a fractured left femur. He also has a hairline fracture of the left tibia. His left leg is supported in what is called a Hamilton-Russell traction system. The system consists of a metal ring which encircles his thigh and a frame to support the lower leg. The whole system and the leg are suspended by a complicated series of ropes and pulleys as shown in Figure 3.1.

FIGURE 3.1 A Hamilton-Russell traction system

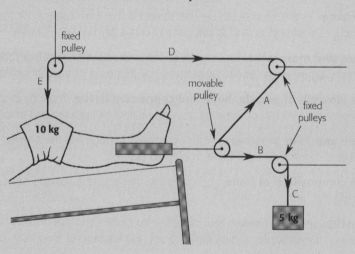

The charge nurse explains the seven main reasons for placing a limb in traction:

1. To immobilise a fracture and maintain alignment.
2. To immobilise and prevent further soft tissue damage.
3. To reduce a dislocated joint.
4. To reduce fractures and realign bone fragments.
5. To rest a diseased joint.
6. To relieve muscle spasm.
7. To correct minor deformities.

In the case of David M., she tells you that one of the ropes, E, is used mainly to support his leg clear of the bed, to prevent pressure sores. The other ropes (A, B, and C) are used to pull his leg to keep the two parts of the fractured femur in alignment and relieve muscle spasm in the thigh muscles surrounding the femur.

Clinical notes for David M.: Nurses are responsible for the following evaluative criteria in such cases: observing for improved body alignment while lying down, sitting or standing; inspecting the skin for possible pressure sores; inspecting the musculoskeletal system for absence of joint contractures.

To understand fully all that has been said about the case of David M., it's necessary to find out what is being done to him and why. This entails close attention to the concepts of force and pressure. Later, we will see that these two concepts can help us understand much more that occurs in the human body, which may influence the way you care for your clients. For example, many explanations given to nurses about traction systems are written in the following way.

The Hamilton-Russell Traction System
The arrangement of the pulleys creates a line of pull on the length of the leg which is twice that on the flexion of the hip, thus providing a traction on the hip approximately two and a half times greater than the actual weight applied. The counter-traction is produced by the incline of the bed with the foot of the bed elevated.

As well as the knowledge of anatomy needed to understand this paragraph (flexion of the hip), there are terms such as 'weight' and 'counter-traction' which come from physics.

3.1 FORCE

The way we commonly use the word 'force' in everyday situations is perfectly adequate for our purposes—for example, pushing heavy furniture or twisting off jar lids. **Force** refers to the way we push or pull or twist something to make it move (though the correct term for twisting is **torque**). When something is moving, it has a speed, often measured in metres per second, or m.s^{-1} (m/sec). If it's not moving, its speed is 0 m.s^{-1}. How much force you end up using depends on the job you need to do and how strong you are and whether or not you're using a machine or a friend to help you.

As you might guess, force is something that we measure. In Chapter 1 we talked about measurement needing a number and a unit. The unit for force is the newton (N), derived from kg, m and sec^{-2} (which means 1/sec^2; see Chapter 1). Can you see how that makes sense? Think again about force as trying to move something. The 'something' you move has a mass, measured in kilograms. Moving something usually means changing its speed, of course, from lying still to moving. Changing speed is called accelerating, as when you accelerate your car or bike. It's measured in m.s^{-2}. Put them together and we have:

$$\text{force} = \text{mass} \times \text{acceleration}$$
$$\text{newton (N)} = \text{kg} \times \text{m.s}^{-2}$$

The acceleration downward, caused by the force of gravity, is about 9.8 m.s^{-2}. This means that every second an object falls, its speed increases by 9.8 m.s^{-1}. A force of 1 N is about the same as the weight of a good-sized apple lying in your hand. In this case, the force is the weight of the object, which is not the same thing as its mass. The mass is a measure of how much matter (atoms, molecules) the object contains, measured in kilograms. Weight is a measure of how strongly that mass is pulled down towards the earth; weight is a force and is therefore measured in newtons. We often confuse these two in ordinary conversation, when we say that a client weighs 60 kilograms. What we really mean is that the client has a mass of 60 kilograms. Their weight is found by using the equation below:

$$\text{weight in newtons} = 60 \text{ kg} \times 9.8 \text{ m.s}^{-2}$$
$$= 588 \text{ N}$$

When you lift clients, you use a force. You have to because they are being pulled down onto the bed by the force of gravity, so it is necessary to pull them

Force
In simple terms, a push or a pull done in an effort to get an object to change its motion. Technically, it's the product of the mass of the object (in kg) and its acceleration (in m.s^{-2}); F = mass × acceleration. It is measured in newtons (N).

Torque
The tendency of a force to produce rotation around a pivot point, as in untwisting a jar lid.

upwards with a greater force. You can do that only if you are strong enough, or have help. Besides being careful not to injure the client, you must take care not to injure yourself by trying to lift, turn, or carry something too heavy for you.

Since the word 'traction' means 'a sustained pull', we now know that the traction system mentioned in the case study is designed to pull on the body with a certain force. 'Counter-traction', therefore, must be a pull in the opposite direction, like the two teams in a tug-of-war.

If the traction and counter-traction have the same force (say, each is 25 N), then technically this is called equilibrium and we can say that 'in equilibrium, the forces are balanced'. Balanced forces add up to zero; see Figure 3.2.

In this tug-of-war, nobody wins. Traction systems like the Hamilton-Russell one mentioned above are always equilibrium systems, holding the client or the broken limb immobile. As the case study indicates, this is one of the important functions of any traction system. Of course, they are also designed to allow as much movement of unaffected areas as possible, for the comfort and better nursing care of the client.

Equilibrium

The concept of **equilibrium** is important; it applies not only to force, but to other situations in the body, such as pressure. Some of these are discussed in later chapters. The key point to remember about force is that it not only has a size (like 2 N or 678 N) but pulls or pushes in a certain direction (such as up, or right to left). Determining the conditions for equilibrium requires an

> **Equilibrium**
> A balance of forces, so that the object does not move, or moves with a steady speed. A client in bed is in equilibrium, balanced by gravity (downwards) and the support of the bed (upwards).

FIGURE 3.2 For equilibrium, the sum of all forces is zero

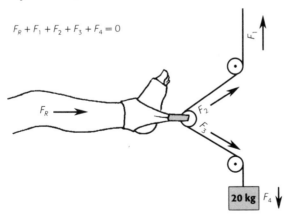

$$F_R + F_1 + F_2 + F_3 + F_4 = 0$$

understanding of the relationship between the sizes and the directions of all the forces in action. This can be done with vectors, as shown in the supplement that follows.

SUPPLEMENT ON VECTORS

We have said that a force has both a size (e.g. 15 N) and a direction (e.g. left to right). Any quantity for which this is true is referred to as a vector. The speed of your car, more properly called its velocity, is also a vector; to specify it completely, you must give both size and direction, say 60 m.s^{-1} north. When more than one vector quantity, such as force, is acting on an object at the same time, what happens? Consider a simple case. David M. is lying in bed. Gravity pulls him downward with a force equal to his weight, say 800 N down. We will refer to this as –800 N to show that it is in a downward direction. He is supported by the bed, which must be pushing upwards with an equal force, 800 N up, which we will call +800 N, as it is in the opposite or upward direction. This is shown in the simple diagram in Figure 3.3.

The resulting force is zero, because

$$(+800 \text{ N}) + (-800 \text{ N}) = 0 \text{ N}$$

and this is equilibrium.

However, consider the situation where David M. is lifted straight upward with a force of 900 N up (+900 N). We need to exert a force equal to his weight (800 N) plus something extra to move him upwards. Since the total force up is +900 N, we have

$$(+900 \text{ N}) + (-800 \text{ N}) = +100 \text{ N}$$

FIGURE 3.3

+800 N (up)

–800 N (down)

The +100 N, or 100 N up, is called the resultant force. It must have both a size and a direction, of course.

A more complicated case arises when the forces, or any vectors, are not in opposite directions. Reconsider the diagram of David M.'s traction system in Figure 3.1 and consider the two ropes labelled A and B. Showing them separate from the traction system, the forces they exert can be represented by two lines, as in Figure 3.4(a).

FIGURE 3.4(A)

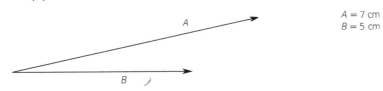

$A = 7$ cm
$B = 5$ cm

We can find the resultant force in two ways if we know the size of the forces *A* and *B*.

Method 1: Completing the parallelogram

To the diagram of Figure 3.4(a) we add two new lines, lines *C* and *D*, which are parallel to and the same length as, *A* and *B*; see Figure 3.4(b).

FIGURE 3.4(B)

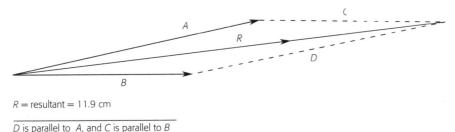

$R = $ resultant $= 11.9$ cm

D is parallel to *A*, and *C* is parallel to *B*

The line marked *R* is the resultant force. Its direction is determined by the diagram and its size is determined by the fact that all lines are to scale, where 1 cm represents 10 N. That is,

A = 7 cm = 70 N
B = 5 cm = 50 N
R = 11.9 cm = 119 N

Method 2: Adding vectors head to tail

We redraw vectors *A* and *B* so that they are now head to tail, as in Figure 3.4(c).

FIGURE 3.4(C)

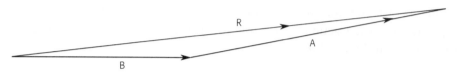

We then draw the resultant vector **R** by connecting the tail of **B**, the first vector we redrew, to the head of the last one we redrew, **A**. Again, the scale gives us its size and the diagram shows us its direction. This method has the advantage that it can be used when a large number of vectors are all acting on the same object at once. We simply scale them, add them head to tail in any order, then connect the resultant from the tail of the first to the head of the last. As you might expect from what you understand about the purpose of David M.'s traction system, the resultant force of all the forces acting is in the direction of his femur, pulling it straight.

3.2 PRESSURE

Pressure
A measure of force per area; P = force/area. It is measured in pascals, though mm Hg is still used in nursing; see Chapter 1.

Force is an important factor in the nurse's concern with pressure sores, or bed sores, a common potential problem with immobilised clients. **Pressure** is simply force/area, the amount of force pushing or pulling on a particular area. Its units are therefore newton/m², written N.m⁻², which is given the name pascal (Pa). Because pressure = force/area, the smaller the area, the bigger the pressure:

$$P = 50 \text{ N}/25 \text{ m}^2 = 2 \text{ Pa; but } P = 50 \text{ N}/5 \text{ m}^2 = 10 \text{ Pa}$$

Think of the weight of David M.'s body as the force pushing on the bed. Places where that force is concentrated on a small area, such as his heels or buttocks, may therefore have a high pressure, perhaps enough to cut off the flow of blood to that area and create a pressure sore.

We also talk a great deal about 'blood pressure'. Here we are talking about the force used to push blood through tubes of different diameters or cross-sectional areas. The idea is the same: the smaller the area, the greater the pressure, unless we reduce the force. This helps to explain why narrowing of the arteries due, for example, to a build-up of fat deposits, can lead to higher blood pressure. More will be said about this aspect of pressure in Chapter 8 on the circulatory system.

Pressure can be used to help diagnose problems. To see how this is done, imagine holding an inflated balloon in your hands. If you squeeze it, the change in pressure is spread evenly throughout the balloon and all the balloon expands slightly. That is, the higher pressure caused by your fingers is transferred everywhere in the balloon, not just where you squeezed. This equal sharing of pressure changes always takes place when the fluid (air or liquid) is trapped in a closed container.

An event that always happens (unless we interfere with the process), is something that scientists, at least, tend to refer to as a principle. In this case, we have **Pascal's Principle**. In more technical terms, it states 'Any change in pressure in a closed vessel is distributed equally everywhere in the vessel'. In nursing, what this can mean is that a rise in pressure somewhere in the body (say, the heart) can be detected somewhere else, in the arm perhaps. That is why taking blood pressure in an artery in the arm is giving you information about pressures at the heart. While the circulatory system is not quite a closed system, Pascal's Principle applies.

> **Pascal's Principle**
> Any change in pressure anywhere in a closed system is transmitted equally to all parts of the system.

A change in pressure anywhere in an enclosed space (perhaps caused by a tumour) can be detected somewhere else, which may be less invasive or more convenient for assessment. For example, a rise in pressure in the fluid surrounding the brain can be detected outside the skull in the fluid around the spinal cord, the cerebrospinal fluid. Figure 3.5 shows this in more detail.

As you can see, Pascal's Principle can be very useful.

FIGURE 3.5 An example of the use of Pascal's Principle in the diagnosis of a brain tumour

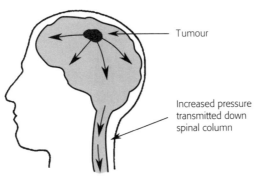

Tumour

Increased pressure transmitted down spinal column

3.3 FORCE AND MOVEMENT

Our client David M. has had his leg X-rayed and the charge nurse lets you have a look at it while explaining how the traction system works and how pressure sores are prevented. The X-ray clearly shows the femur, knee joint, tibia, fibula and (faintly) surrounding muscle. The question you're now asked is, 'How do these structures work together to provide movement?' Figure 3.6 shows the bone, joint and muscle system in the arm with greater clarity than an X-ray would.

The muscle tissue is attached to the bone by tendons, which (along with bone) consist of connective tissue. As you might expect from the name, connective tissue connects things together. When the muscle contracts, it pulls on the bone it's attached to. Because that bone is joined to other bones at the joint, it cannot move freely, but only in certain directions. The best way to understand this may be to examine the way a chicken leg (upper and lower leg, connected at the knee) or wing moves (before you cook it is best). You could examine your own leg, but the muscles may not be easily seen. Take the skin off the chicken leg and look for muscles attached to bone by tendons. Then move the leg as if the chicken were walking. Some movements are easy, because of the way the knee is built. Other movements are clearly more difficult. Refer to your anatomy text and try to picture the various movements discussed there, such as extension, flexion, rotation and so on, in terms of the interaction between the muscles and bones of the chicken leg.

The movements of the limbs (e.g. the knee and lower leg) are often described as if they were **levers**. Because you are likely to hear the lower leg and knee referred to as a 'third-class lever' in your nursing practice, you should have some idea of what that means. Consider Figures 3.7(a)–(d).

The crowbar is called a first-class lever. The name refers to the arrangement of its parts (force, fulcrum and load) and applies to any lever with this arrangement.

Lever

A combination of load, force and fulcrum. Levers either allow greater strength (first and second class), or a greater range of movement (third class)

FIGURE 3.6 Bone, joint and muscle arrangement in the human arm

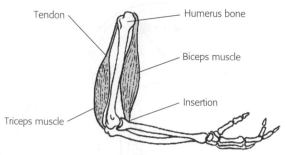

Source: *Physics: Principles with Applications*, 3rd Ed, Giancoli, D.C., 1991, p. 212. © Reprinted by permission of Pearson Education Inc., Upper Saddle River, N.J.

FIGURE 3.7(A) First-class lever

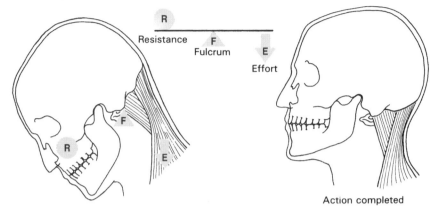

Source: *Fundamentals of Anatomy and Physiology*, 2nd Ed, Martini, F., 1992, p 331. © Reprinted by permission of
Pearson Education Inc., Upper Saddle River, N.J.

FIGURE 3.7(B) Second-class lever

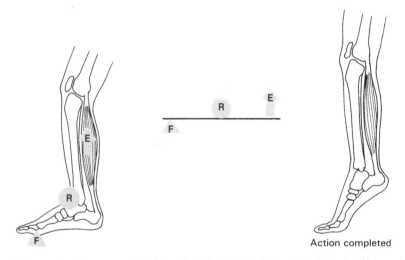

Source: *Fundamentals of Anatomy and Physiology*, 2nd Ed, Martini, F., 1992, p. 331. © Reprinted by permission of
Pearson Education Inc., Upper Saddle River, N.J.

In the body, the force is caused by the contraction of a muscle, or group of muscles
acting together. The fulcrum, or pivot point, is the joint such as the upper vertebra
of the neck when we nod our heads. The load is whatever the muscle pulls and
moves; for nodding, the load is the head. In most cases, the load is the weight of
the limb the muscles move plus any extra weight the limb is holding, such as a
brick; see Figure 3.7(a). As it happens, there are very few first-class levers in the
human body. Nodding the head is a good example of one, however.

FIGURE 3.7(C) How a third-class lever is arranged

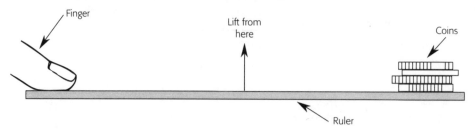

Finger

Lift from
here

Coins

Ruler

FIGURE 3.7(D) Third-class lever in the arm

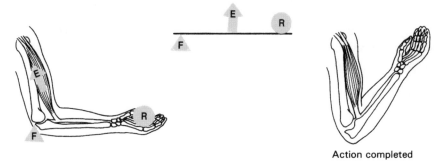

Action completed

Source: *Fundamentals of Anatomy and Physiology*, 2nd Ed, Martini, F., 1992, p. 331. © Reprinted by permission of Pearson Education Inc., Upper Saddle River, N.J.

Second-class levers have the parts arranged differently: fulcrum, load, force. The best example of this is a wheelbarrow. Again, there are few of these in the body, but Figure 3.7(b) shows an example. When standing on tip-toe, the fulcrum is at the point where the toes touch the ground. The load is the weight of body supported by the bones of the foot and the effort is applied by the Achilles tendon.

The most common lever system in humans is the third-class lever. Here, the arrangement is: fulcrum, force, load. To visualise this one, place a ruler flat on the table and hold one end down so it acts like a door hinge. That end is now the fulcrum. Place a couple of coins right at the other end of the ruler. That end is now the load. Now lift the ruler from the centre, as shown in Figure 3.7(c).

Believe it or not, this unusual way to lift things is the one our joints use almost all the time. Figure 3.7(d) shows an example of how the third-class lever works with the arm.

For the body, the advantages of the third-class lever system, as opposed to first and second class, are speed and dexterity of movement. The muscle causing the force has to contract only a short distance and as a result the arm moves a greater distance and at a greater speed. This is discussed in more detail below.

Physicists have the answer to why first-class and second-class levers make you stronger and third-class levers make you faster. Imagine a simple seesaw lever system as shown in Figure 3.8(a), with a load on either end.

When the seesaw is balanced on the fulcrum, the force on one side of the seesaw is the same as on the other. That is, $F_1 = F_2$ as long as the distances from the fulcrum are the same, as they are in this case.

As you know, the position of the load is important. If we slide load A to the end of the seesaw, there will no longer be balance. This relationship between force and position is explained in terms of torque (T), which is defined as:

torque = force × distance from fulcrum;

in symbols, $T = F \times D$.

FIGURE 3.8(A) Equilibrium

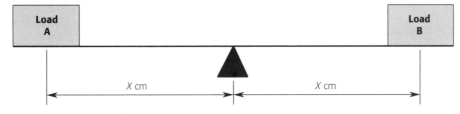

FIGURE 3.8(B)

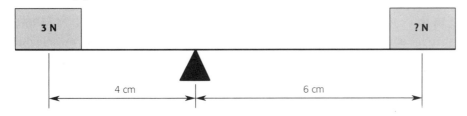

FIGURE 3.8(C)

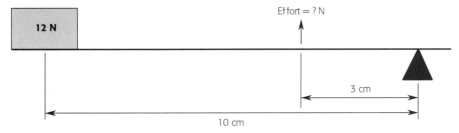

For our seesaw, when it's in balance, we have:

$$T_1 = F_1 \times D_1 \text{ and } T_2 = F_2 \times D_2 \text{ and } T_1 = T_2$$

Now look at the seesaw in Figure 3.8(b) and we'll put in some numbers this time.

Left side (load)	Right side (effort)
$T_1 = 3 \text{ N} \times 4 \text{ cm}$	$T_2 = ? \text{ N} \times 6 \text{ cm}$
$= 12 \text{ N.cm}$	

$$12 \text{ N.cm} = ? \text{ N} \times 6 \text{ cm}$$

Using algebra, $? = 12 \text{ N.cm}/6 \text{ cm} = 2 \text{ N}$

This means that in order to lift something weighing 3 N, you need only 2 N of force. However, since you're further from the fulcrum than the load is, you have to move your side of the seesaw a distance downwards that is greater than the distance that the load moves upward. Try it and see, using the ruler and coins.

The situation for our third-class levers is illustrated by Figure 3.8(c).

Left side (load)	Middle (effort)
$T_1 = 12 \text{ N} \times 10 \text{ cm}$	$T_2 = ? \text{ N} \times 3 \text{ cm}$
$= 120 \text{ N.cm}$	

$$120 \text{ N.cm} = ? \text{ N} \times 3 \text{ cm}$$

In this case, $? = 120 \text{ N.cm}/3 \text{ cm} = 40 \text{ N}$. Here we need to use 40 N of force just to lift something weighing only 12 N; but we have to move the muscle only a little way and the load moves much further. Again, try it with the ruler.

One implication of this for nurses is in the lifting of heavy objects. Hold the object so it lies as close to the fulcrum as possible. In lifting, the fulcrum is most likely to be the elbow or the wrist. In the example at Figure 3.8(c), reducing the load's distance from 10 cm to 5 cm would reduce the force needed to lift it from 40 N to 20 N. Can you do the calculation to show that this is true?

Centre of gravity

You may already have had some experience with assisting clients from bed, or into chairs. Care must be taken that they (and you) do not topple over. In more technical terms, we are concerned that they remain in stable equilibrium. Every object has a point, called the centre of gravity, around which opposing torques are equal. Opposing torques try to rotate the object in opposite direction, say clockwise and anti-clockwise. If we can support an object under its centre of gravity, it will be balanced, or in stable equilibrium. Naturally, concerns for client comfort and safety are paramount, but lifting should be done by support under the centre of gravity whenever possible.

3.4 THE STRENGTH OF BONE, MUSCLE AND CONNECTIVE TISSUE

The traction system set up for David M. uses balanced forces to immobilise his leg and aid the healing process. But of course the reason he's in hospital in the first place is because force broke that same bone. How strong are bones? The answer depends on the way the bone is broken and the type of bone being tested. You can get some idea of bone strength by trying to break some fresh chicken bones; leg bones work well. Keep in mind that there are several different ways to go about breaking a bone. They can be snapped, twisted, pulled apart or crushed. If you try this, a little experimentation would show that bone is generally easiest to break by twisting and hardest by pulling or crushing. The resulting break, in all cases, is referred to as a fracture. There are at least 17 different types of fracture, depending on how the bone is broken and what the resulting damage is. For example, a spiral fracture often results when a bone is twisted and an impacted fracture is commonly caused by crushing. See Figure 3.9 for a classification of fractures.

The more precise measurement of bone strength is done in the same way that steel or wire or any other solid is tested. It is clamped into a machine and a breaking force is applied. Table 3.1 compares bone, human cartilage and other tissues with common materials.

Several interesting things can be determined from Table 3.1. Note that muscle is much weaker than tendon. This helps to make sense of the fact that muscles usually have a greater mass than tendons and that they can tear more easily. Note too that as people age their bones become more brittle. Table 3.1 shows that elderly bone weakens to about the same strength as tendon. It has been claimed that some elderly people, if alarmed or frightened, may violently contract a muscle with enough force to snap off the head of the femur. The resulting fall may cause further damage.

FIGURE 3.9 A classification of fractures

Fracture type

Closed, or *simple*, fractures are completely internal; they do not involve a break in the skin. *Open*, or *compound*, fractures project through the skin. They are more dangerous because of the possibility of infection or uncontrolled bleeding.

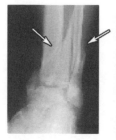

A *Pott's* fracture occurs at the ankle and affects both bones of the lower leg.

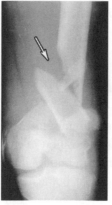

Comminuted fractures shatter the affected area into a multitude of bony fragments.

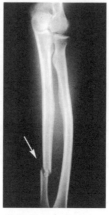

Transverse fractures break a shaft bone across its long axis.

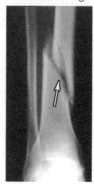

Spiral fractures, produced by twisting stresses, spread along the length of the bone.

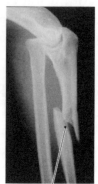

Nondisplaced fractures retain the normal alignment of the bone elements or fragments. *Displaced* fractures produce new and abnormal arrangements of bony elements.

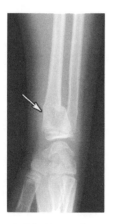

A *Colles'* fracture is a break in the distal portion of the radius, the slender bone of the forearm; it is often the result of reaching out to cushion a fall

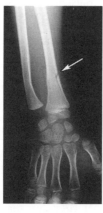

In a *greenstick* fracture only one side of the shaft is broken, and the other is bent; this usually occurs in children whose long bones have yet to fully ossify.

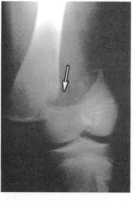

Epiphyseal fractures usually occur where the matrix is undergoing calcification and chondrocytes are dying. A clean transverse fracture along this line usually heals well. Fractures between the epiphysis and the epiphyseal plate can permanently halt further longitudinal growth unless carefully treated; often surgery is required.

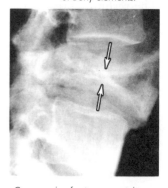

Compression fractures occur in vertebrae subjected to extreme stress, as when landing on your seat after a fall

Source: *Fundamentals of Anatomy and Physiology*, 5th Ed, Martini et al, 2001, p. 186. © Reprinted by permission of Pearson Education Inc., Upper Saddle River, N.J.

TABLE 3.1 STRENGTH OF MATERIALS

Material	Breaking strength in MN.m^{-2}
Muscle	0.1
Cartilage	3
Ordinary brick	7
Skin	10.3
Tendon	82
Bone (young adult)	110
Bone (elderly)	85
Glass	35–175
Nylon thread	1050
Steel wire	3100

3.5 BONE AND JOINT PROSTHESES

In the orthopaedics ward where David M. is, there are likely to be clients who have had hip replacement surgery. If you can, examine a modern hip replacement prosthesis. Many of these are made of two or more different materials joined together. Today, that usually means a titanium metal shaft and a plastic or ceramic head; see Figure 3.10 for some examples.

If they are going to take the place of living bone, these prostheses must act as much like bone as possible. They must not be too stiff or too flexible, they must not be toxic or easily corroded, they must not be electrically active and they must be easily connected to living bone. This matching becomes more complicated when you think about people getting older.

Bone is elastic: it deforms slightly when loaded or stressed and then returns to its original shape when unloaded. The cement used to fix the prosthesis does not adhere to bone, rather it keys into the bone so that the implant fits accurately the cavity of the living bone marrow at the time of the operation. The resulting metal/cement mixture is also elastic. Each component, however, has a different stiffness . . . Bone, being a living, dynamic material, attempts to re-orientate itself to the new patterns of stress and strain. At the same time, the blood supply to the bone, which was damaged when the surgeon inserted the implant, also has to be re-established.

Source: *New Scientist*, 10 Dec. 1987, p. 36.

The problems don't end there. The materials of which the prosthesis are made are chemically different from bone. The nurse must be aware of the

FIGURE 3.10 Examples of artificial joints

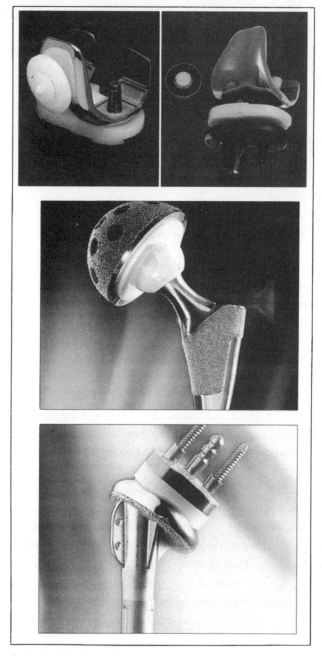

Source: *Fundamentals of Anatomy and Physiology*, 2nd Ed, Martini, F., 1992, p. 229. © Reprinted by permission of Pearson Education Inc., Upper Saddle River, N.J.

possibility of these differences because they can lead to slower healing and increase the risk of infection. In 1987 it was estimated that there were 400 000 hip replacements done worldwide. If only 1% became infected, that means an extra 4000 referrals a year.

3.6 THE COMPOSITION OF BONE

As it is now clear that the function of bone depends on the structure of bone, it's important that we spend a little time looking at what bone is made of.

Bone is a composite material. It has long organic fibres (i.e. made from living cells in the bone) made of collagen and these fibres are cemented with water and inorganic crystals. The crystals are made from minerals and make up 70% of the bone.

What do we mean by minerals in the bone? Let's look at them first. Even before birth, the collagen fibres in the foetus are being strengthened by a cement containing lime, phosphorus and water. The symbols for these substances are: lime, composed of calcium (Ca), hydrogen (H), and oxygen (O); and phosphorus (P). These are naturally occurring substances, taken in with food, air and water. Once in the body, they are deposited on and between the collagen fibres, forming a hard, rock-like matrix that surrounds the collagen and strengthens it.

Much more will be said about how these substances get into the body and what they do when we discuss nutrition in Chapters 4 and 5. For the moment, it's only important that you notice what substances are present in bone.

One of these minerals, calcium, is often talked about. It is found in milk, for example, and is essential for building bone. If it is lacking in the diet, or lost from the body, the bones become brittle and fracture more easily. This loss of calcium contributes to the condition referred to as osteoporosis. It occurs more frequently in elderly people in general and postmenopausal women in particular. As many as 25% of women develop this condition.

Part of the prevention of osteoporosis may be a diet rich in calcium, ideally from early in life. Oestrogen therapy is another, more recent possibility, but is beyond the scope of this book.

3.7 MUSCLES AND MOVEMENT

If we think of a bone as the lever and a joint as the fulcrum, then a muscle supplies the force. Muscles work by contracting, shortening their length and pulling on the bone to which they are attached. An anatomy and physiology textbook can give you the details of muscle contraction, though it is outlined later in this chapter. At this point we shall focus on **work** and **energy**, because we commonly talk about muscles using energy and doing work, and it is important to understand exactly what is involved.

Work and energy are closely related. When a muscle pulls a bone, two things are involved. First, as the muscle tenses, it pulls with a certain force, which may be strong or weak. Next, the muscle contracts, shortening by a certain distance, small or larger. If we multiply the force, F, by the distance moved, d, we have the work done by the muscle on the bone.

Work

The product of force times distance moved. The work done by a muscle, for example, is the force applied by the muscle times the distance it moves. W = force × distance. It is measured in joules.

$$\text{Work} = \text{force} \times \text{distance}$$

In symbols, $W = F \times d$. Remembering that force is measured in newtons (N) and distance in metres (m), work is measured in $N \times m$ or N.m. This is a derived unit, given the name joule (J). Both work and energy are measured in joules. For example, if the muscle pulls with a force of 0.1 N and contracts by 0.05 m, then $W = 0.1 \text{ N} \times 0.05 \text{ m} = 0.005 \text{ J}$. We say that the work is done by the muscle on the bone.

To do work, the muscle needs energy. In fact, energy has often been defined as 'the ability to do work'. You might find it helpful to think of energy as causing change. For every type of change, there is a type of energy. For example, if something gets hotter or colder, we talk of heat energy. If it becomes electrically charged, we speak of electrical energy.

Energy

Often defined as the ability to do work and responsible for causing change. It is measured in joules. There are many types, but the two basic forms are kinetic energy and potential energy.

Two types of energy are especially important for muscles. When something changes its speed by moving, as a muscle does when it contracts, we talk of **kinetic energy**. Kinetic means motion. Anything in motion is said to have kinetic energy. How much kinetic energy it has depends on how great its mass is and how fast it's moving. Technically,

Kinetic energy

The energy of objects in motion. It is found by
$$E_{kinetic} = 1/2 \times \text{mass} \times \text{velocity}^2$$

$$E_{kinetic} = 1/2 \text{ mass} \times \text{velocity}^2$$

The other main type is called chemical energy. It can be defined as the energy stored in a muscle in the form of energy-rich molecules called ATP, discussed in Chapter 5. These chemicals, made by the body from the nutrients derived from our diet, are formed during the process of cellular respiration, which is also discussed in greater detail in Chapter 5.

Chemical energy is stored up in the muscle, ready to be used to allow the

Potential energy

The energy of position, usually taken to mean how high from the ground, or how far apart. It's found by
$$E_{potential} = \text{mass} \times \text{acceleration} \times \text{position}$$

muscle to contract. An example of chemical energy that may be more familiar to you is a battery for a radio or torch. The chemical energy is stored in there, waiting for you to use it. In a muscle, the energy is waiting for you, unless it's exhausted from overuse.

We can change one type of energy into another quite easily. Electrical energy is changed partly to light energy and partly to heat energy in a light globe and some of the chemical energy stored in petrol is changed to kinetic energy for a car. In our muscles, chemical energy is changed to kinetic energy and heat energy whenever we wish. Sometimes this happens without us having to think about it, as in the beat of the heart, or the involuntary blinking of our eyes.

Remember what was said about the connection between structure and function? The function of the muscle is to contract and it uses its energy to do just that. The structure of the muscle makes this contraction possible.

Muscle tissue is composed of myriads of fibres, arranged so they lie parallel to one another along the length of the muscle. When these fibres shorten, the muscle contracts. Although the details of the contraction of the fibres are complex, the nurse should begin to understand the need for proper diet to supply energy for muscles, the role of the muscle in pulling by contraction and the connections between muscle and bone that allow movement.

3.8 ENERGY IN THE BODY

If energy allows us to do work, then it's obviously needed not just in muscle but everywhere in the body. To maintain homeostasis, for example, the body continually uses energy. In fact, one definition of 'life' is the ability to use energy to maintain structure and function. First of all, think of the body as a whole. There are inputs and outputs, as shown in Figure 3.11.

Some of the energy that enters the body is changed into forms useful for

FIGURE 3.11 Energy inputs and outputs

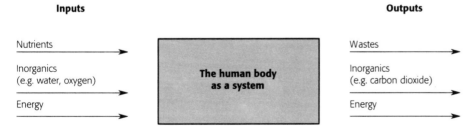

maintaining homeostasis and for growth. The energy that leaves the body is usually in a form the body cannot easily use, such as heat.

If we could add up the number of joules of energy the body takes in, we would find that it equals the amount used plus the amount given out. This is an example of the law in physics that energy cannot be created or destroyed. It can only change form. For example, we change the chemical energy stored in an ATP molecule into a wide variety of forms, such as mechanical. As you know, the body cannot turn all the incoming energy into useful energy. Some is always wasted, usually as heat, though we use some of that energy for maintaining a constant body temperature, as already mentioned. Though it's hard to calculate this very precisely, the body has been said to be anywhere from 30% to 40% efficient, which is close to twice as good as an automobile.

In this chapter we have discussed energy in only the broadest possible way. The details of energy gain and use by the body forms an important part of Chapter 5, on nutrition, and you may wish to turn to those pages now. The muscular/skeletal system, in collaboration with the nervous system, is a major user of energy for motion and handling objects. Through it, we act on the world.

Questions

Level 1

1. Define the following: force, vector, resultant, pressure, work, energy.
2. What is meant by equilibrium of forces?
3. Distinguish between force and work. If you hold a chart stationary in your hands, are you exerting a force? Are you doing work on the chart?
4. How much force is being exerted when a mass of 25 kg is accelerated at 2.5 m.sec^{-2}?

   Answer: 62.5 N
5. What are some of the components of bone and what are some of its physical properties?

Level 2

6. How does the Hamilton-Russell traction system keep David M.'s leg in equilibrium?
7. Explain why bed sores occur on the heels or buttocks.
8. How can Pascal's Principle be used to diagnose a tumour in the eye?
9. Use a diagram to explain the differences between the three types of levers. How does the elbow act as a third-class lever?

Level 3

10. Using the definition of energy given in the glossary at the end of this book, explain why there must be several different types of energy.
11. Discuss the efficiency of the human body in the way it uses energy. Do you think 30–40% efficiency is very good? Do you think professional athletes are more efficient?
12. Why do you think we have come to use third-class levers for most of our joints?

CALCULATION PROBLEM RELATED TO THE CASE STUDY

David M. has a mass of 90 kg. Assume that, as he lies in bed, 20% of that mass is resting on his shoulder blades, which have a combined area of 110 cm². What is the pressure (in mm Hg) at each shoulder blade?

Answer:

20% of 90 kg = 20/100 × 90 kg = 18 kg
Force = $m \times a$ = 18 kg × 9.8 m.s^{-2} = 176.4 N
Pressure = F/A = 176.4 N/110 cm² = 176.4 N/0.011 m² = 1.6 × 10⁴ Pa
Since 1 kPa = 7.5 mm Hg, then 16 kPa = (16 × 7.5) mm Hg = 120 mm Hg.

THE GENERAL WARD AND CLIENT CARE: PART 1

ATOMS, MOLECULES AND CHEMICAL BONDS

Chapter outline

4.1 **What is a chemical?** Your objective: to be able to provide definitions and examples of atoms, ions, elements, molecules and compounds.

4.2 **Bonding.** Your objective: to understand in general terms how molecules are held together, including the idea of valency.

4.3 **Organic compounds in the body.** Your objective: to be able to describe the structure and function of carbohydrates, fats, proteins and nucleic acids.

4.4 **Functional groups.** Your objective: to know the major groups of specific atoms that determine the name and function of important organic compounds.

4.5 **Chemical homeostasis.** Your objective: to understand the importance of maintaining the right number and type of chemicals in the body.

INTRODUCTION

This chapter and the next provide you with some beginning chemistry. This first chapter contains information on atoms, ions and molecules, the types of molecules commonly found in the body and how they combine to form substances essential for proper body functioning.

Case Study: Mrs Laurel S.

Laurel S. is in her late 50s. On admission she had a weight of 175 kg. She has a history of adult-onset diabetes. She is in hospital recovering from a septic ulcer on her left foot. A glucose tolerance test administered on admission indicated a level of glucose in the blood of over 150 mg/mL, indicating loss of insulin control. One reason for this seems to be her poor attention to diet. On her chart is pinned the following information, placed there for reference by student nurses:

Adult-onset diabetes in which endogenous insulin is ineffective for the body demands. Heredity, ageing, lifestyle, certain medical disorders and some drugs influence its development. Approximately 80% of adult-onset diabetics are obese at the time of diagnosis. Treatment is usually by diet and hypoglycaemic drugs or by diet alone.

Source: Commonwealth Dept. of Health (1980) *Hospital Diet Manual*, AGPS, Canberra, p.55.

Clinical note for Laurel S.: Nursing care for this patient would involve continuous monitoring of such chemistry-related values of serum osmolality, electrolytes, glucose, haemoglobin, haematocrit and urine ketones. Also, there may be a need to administer prescribed regular insulin when serum glucose falls in the range of 250–800 mg/dL.

It seems clear from our case study that, in order to understand what is wrong with Laurel S. and why she is being treated this way, you must know some chemistry. Many beginning nursing students become quite alarmed at hearing this, perhaps because they were given little or no chemistry in their schooling, or they may have come to think of chemistry as very difficult and too complicated. Hopefully, by starting at the very beginning, this chapter will show that chemistry can be both useful and easier to understand than you may have feared. If you have some experience with chemistry, this chapter will be both a helpful review and an opportunity to see chemistry applied more specifically to a nursing context.

4.1 WHAT IS A CHEMICAL?

Any chemistry book will tell you that chemicals are the substances that make up the world. All the objects for which we have common names—rocks, water, air, and so on—are made of chemicals. Since our goal is to understand the structure and function of the human body, we need to know something of the chemicals of which it is made. The approach we will take is to start by looking at atoms, of which chemicals are composed. From atoms we can then discuss ions, followed by the combinations of atoms called molecules and finally something of the forces that hold these particles together. Figure 4.1 shows a common way of classifying substances used by chemists.

Atoms

Chemicals are made of **atoms**. An atom is defined as the smallest portion of an element (such as iron or oxygen) that cannot be broken down to anything smaller

Atom

The smallest unit of an element. Since there are 92 naturally occurring elements, there are 92 naturally occurring types of atoms.

FIGURE 4.1 A diagram showing the classification of substances.

'Separation by physical means' involves using differences in the physical properties of substances to separate them. 'Separation by chemical means' involves transforming a substance into one or more other substances.

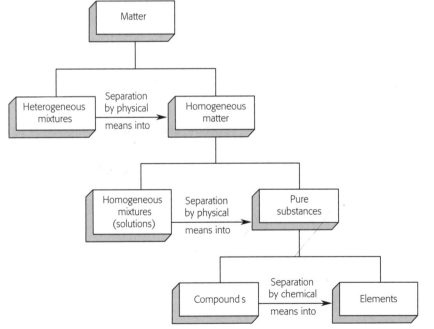

Source: *Chemistry; The Central Science*, Brown, T.L. and Le May, H.E., 1991, p. 22. © Reprinted by permission of Pearson Education Inc., Upper Saddle River, N.J.

by chemical means. The whole staggeringly complex world is made of only about 92 different types of atoms, one type for each of the elements. We say 'about 92' because there are 92 elements that occur naturally on earth and another dozen or so that have been made by humans in small quantities. No matter what substance you think of, it is made from some combination of a limited number of different kinds of atoms. These atoms are very small, around 10^{-10} m across. As a result, one hundred trillion of them (100 000 000 000 000 or 10^{14}) could be placed on the head of a pin. For a model of the structure of one of these atoms, see Figure 4.2.

The central **nucleus** has two types of particles in it, called protons and neutrons. Surrounding the nucleus, and moving around it like the moon around the earth, are the **electrons**. In terms of size, the protons and neutrons are equal to each other, but both are almost 2000 times larger than electrons. The distance between the nucleus and the electrons, compared to the size of an atom, is also very large, just as the moon is a great distance away from the earth. The electrons are found to orbit the nucleus only at specific distances and only fixed numbers of electrons are found in any one orbit.

What holds an atom together? Remember our definition of force as a push or pull? Well, there is an **electrostatic force** that pulls the electrons and protons towards each other. The electron has an electric charge, which is conventionally termed negative. A proton has an electric charge referred to as positive. Positive and negative charges are opposites. It is misleading to think of a negative charge as being something less than zero. A better idea is that of black and white. These opposite charges attract one another and we refer to this as an electrostatic force of attraction. This attraction is similar to the north pole of a magnet attracting the south pole of another magnet.

Nucleus

The central region of an atom, consisting of one or more protons (with a +1 charge) and one or more neutrons (with no charge).

Electron

The particle that circles the nucleus. Much smaller than either the proton or neutron, it has a −1 charge.

Electrostatic force

The force of attraction and/or repulsion between electric charges. Opposite charges attract (+ to −) and like charges repel. It holds the electrons to the nucleus and holds ions of different charges together in ionic bonds.

FIGURE 4.2 Simple model of atomic structure; average atomic diameter is 10^{-10} m

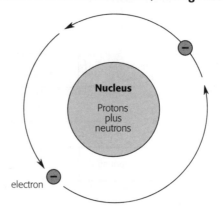

Particles with the same charge will push or repel each other, just as two north magnetic poles will. That means that two electrons, or two protons, will repel each other. (Protons are held together in the nucleus by another force, called the strong nuclear force, discussed in Chapter 12.) The neutron has no charge and is therefore neutral. This is summarised in Table 4.1.

The various types of atoms differ in the number of protons, neutrons and electrons they contain. For instance, there is an atom with only one proton in the nucleus, no neutrons and 1 electron circling the proton. It's called hydrogen. Another atom, with two protons and two neutrons in the nucleus, with two electrons outside, is called helium. The atom with eight protons, eight neutrons and eight electrons, is called oxygen. (Some oxygen atoms have more or less than eight neutrons in the nucleus, forming isotopes of oxygen. Most atoms form isotopes, which are discussed in Chapter 12.) If we go all the way to the atom with 92 protons, we find it has about 142 neutrons and 92 electrons and is called uranium. Perhaps you've noticed that the number of protons in these examples has always been the same as the numbers of electrons. These atoms are said to be electrically neutral. That is, if we add the number of positive charges (the protons) and the number of negative charges (the electrons) the answer is zero. For example:

Hydrogen:	1 proton	+	1 electron		
	+1	+	−1	=	0
Helium:	2 protons	+	2 electrons		
	+2	+	−2	=	0
Oxygen:	8 protons	+	8 electrons		
	+8	+	−8	=	0

Ions

It is possible for an atom to lose one or more of its electrons. For example, they can be 'rubbed off' or 'stolen' by other atoms. When this happens, the atom is

TABLE 4.1 STRUCTURE OF THE ATOM

Particle	Size (m)	Charge
Proton	10^{-15}	+1
Neutron	10^{-15}	0
Electron	10^{-18}	−1

Ion

An atom or molecule that has either gained or lost one or more electrons. If it has gained, it is referred to as a negative ion; if lost, it's a positive ion.

no longer neutral. It is now called an **ion**. For example, if hydrogen loses its electron, then

Hydrogen:	1 proton	+	0 electrons	
H^+	+1	+	0	= +1

The hydrogen atom is called a positive ion, or cation and has a charge of +1. Other examples are:

Calcium ion:	20 protons	+	18 electrons	(lost 2)
Ca^{2+}	+20	+	−18	= +2
Sodium ion:	11 protons	+	10 electrons	(lost 1)
Na^+	+11	+	−10	= +1

Of course, if one atom 'steals' electrons from another, then it has too many to be neutral any more. It is now a negative ion, or anion.

Here are two more examples. Notice that sometimes, when an atom becomes an ion, its name changes slightly. So, instead of a chlorine atom, we have a chloride ion.

Chloride ion:	17 protons	+	18 electrons	(gains 1)
Cl^-	+17	+	−18	= −1
Iodide ion:	53 protons	+	54 electrons	(gains 1)
I^-	+53	+	−54	= −1

The main thing to remember as a nurse is: an atom is neutral and an ion is charged. Atoms become ions when they lose or gain one or more of their electrons. Ions common in the human body are listed in Table 4.2.

Note that an atom will usually either gain or lose electrons, though some atoms can do either. Chemists use this fact as part of their classification of the elements into types, or groups. One of the characteristics of metals like iron or silver, for example, is that they always lose electrons. Non-metals, such as chlorine or iodine above, always gain them. While other atoms can lose or gain electrons depending on circumstances, they usually have a tendency for one or the other. The explanation for this lies with the valency, to be discussed later in this chapter.

Unless they are radioactive, atoms never lose or gain any of their protons or neutrons, as these are locked inside the nucleus. It is the number of protons that determines what type of element it is, not the number of electrons. The number of protons is called the atomic number of the element and, as Figure 4.3 shows, it is followed by the name and symbol of the element on the Periodic Table.

TABLE 4.2 COMMON IONS IN THE HUMAN BODY

Ion	Location	Function	Excess	Deficit
Na^+	Usually outside cell	Control of body fluids	Hypernatraemia, thirst, oedema	Hyponatraemia, diarrhoea, decrease in body fluids
K^+	Usually inside cell	Control of body fluid and cell function	Hyperkalaemia, irritability, cardiac arrest	Hypokalaemia, lethargy, muscle weakness
Cl^-	Usually outside cell	Control of fluids	Same as Na^+	Same as Na^+
Ca^{2+}	In bone	Muscle function	Hypercalcaemia, relaxed muscles	Hypocalcaemia cramps, tetany
Mg^{2+}	In bone	Muscle & nerve function	Drowsiness	Hypertension

FIGURE 4.3 Some useful information from the Periodic Table

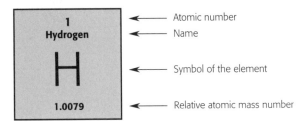

Elements

So far we have used the word 'atom' when talking about chemicals and we have said there are about 92 different types of atoms. An **element** is composed of atoms all of the same type. We have already used the names of some of the elements above. Hydrogen is an element, for example. A bottle of hydrogen would contain only one type of atom and of course they are all hydrogen atoms. Carbon is an element, so pure carbon contains only carbon atoms. Atoms have often been referred to as the building blocks of the elements and so all the substances in the universe are made up of only these 92 or so building blocks. As it happens, only about 10 of the 92 elements are commonly found in the human body, another eight are required by the body in small amounts and a few more in trace amounts.

Element

A substance that cannot be broken down into any simpler substance. There are 92 naturally occurring elements, about 16 of which are crucial for healthy body function.

TABLE 4.3 MAIN ELEMENTS REQUIRED BY THE HUMAN BODY

Name	Symbol	Symbol of ion (if found)
Carbon	C	
Hydrogen	H	H^+
Oxygen	O	O^{2-}
Nitrogen	N	
Iron	Fe	Fe^{2+} or Fe^{3+}
Calcium	Ca	Ca^{2+}
Phosphorus	P	
Potassium	K	K^+
Sodium	Na	Na^+
Sulphur	S	S^{2-}
Necessary in small or trace amounts		
Chlorine	Cl	Cl^-
Cobalt	Co	Co^{2+}
Fluorine	F	F^-
Iodine	I	I^-
Magnesium	Mg	Mg^{2+}
Zinc	Zn	Zn^{2+}
Copper	Cu	Cu^{2+}
Selenium	Se	
Manganese	Mn	
Molybdenum	Mo	

Try to memorise at least the first 10 elements in Table 4.3, as well as chlorine and iodine and their symbols.

The full list of elements and symbols appears in the Periodic Table, where they are arranged in order of increasing mass. The rows and columns organise them into groups with similar chemical properties.

Molecules

If we could see things as small as these, we would see that a block of pure gold, which is an element, has only single gold atoms in it and a single gold atom is symbolised as Au. But a bottle of oxygen, which is also an element, has pairs of oxygen atoms in it. When these combinations of atoms occur, the symbol for the element includes a subscript that shows how many atoms are joined together—

for example, O_2, H_2 and I_2. We refer to these combinations as **molecules**, which means two or more atoms joined together. So if someone mentions an atom of oxygen, they mean only one. But if they talk about a molecule of oxygen, they mean O_2.

Molecules, like atoms, can form ions, which means that one or more electrons have been gained or lost by the molecule as a whole—for example, an hydroxide ion, OH^-, or a carbonate ion, CO_3^{2-}.

Molecule
One or more atoms joined together chemically; for example, H_2 and CO_2.

Compounds

One thing we know for sure is that there are a lot more than 92 different types of objects in the world, which means that the 92 atoms of the elements must combine together in various ways to make them all. When they do combine, the result is called a **compound**. Quite simply, a compound is a combination of two or more atoms of different types. Most compounds are, therefore, also molecules. If this sounds confusing, think of it this way. If all the atoms in the molecule are the same type, it is a molecule of an element. For example, we can have a molecule of oxygen, O_2. But if the atoms are of different types and they are chemically joined to make a molecule, then it is a molecule of a compound. For example, carbon dioxide, CO_2, is a compound made of molecules of that type.

A compound is written by using the symbols of the elements that are in it, plus a subscript to tell you how many atoms of that element there are. See Table 4.4 for a few examples.

The human body is made of thousands of different compounds. Many of them are very complicated, but some are as simple as water or carbon dioxide.

Compound
One or more atoms of *different* types joined together chemically. Strictly speaking, the atoms are joined in a fixed proportion of one to another. CO_2 is a molecule of a compound; H_2 is a molecule of an element.

Mixtures

One question you may have been asking yourself is whether air, for example, is a compound. The easy answer is no. Air is a **mixture** of different molecules,

Mixture
A grouping together of elements, compounds and/or molecules which are *not* chemically joined by bonds and are not joined in a fixed proportion; for example, air, or a mixture of sand and salt.

TABLE 4.4 EXAMPLES OF COMPOUNDS

Name of substance	Symbol	What it means
Water	H_2O	2 hydrogen atoms, 1 oxygen atom
Table salt	NaCl	1 sodium ion, 1 chloride ion
Carbon dioxide	CO_2	1 carbon atom, 2 oxygen atoms
Glucose	$C_6H_{12}O_6$	6 carbon, 12 hydrogen, 6 oxygen

such as O_2, CO_2 and N_2. These molecules are separate from each other in the air. A mixture, then, contains different molecules that are not chemically combined. Think for a moment of a house; it's a mixture of brick, wood, glass, metal and many other things. Similarly, there are lots of mixtures in the body, which we will be talking about as we go along. For example, blood is one very important mixture of water, salt, oxygen, blood cells and many other substances.

Chemical reactions

It is clear that a compound consists of two or more elements joined together, and this is shown by a chemical equation. An equation is a shorthand way of representing a chemical reaction. It shows what happens when one or more chemicals, called reactants, undergo chemical changes to form new chemicals, called products. For example, in words:

reactant 1 + reactant 2 → product 1 + product 2

Examples

$$C + O_2 \rightarrow CO_2$$
$$2H_2 + O_2 \rightarrow 2H_2O$$

A slightly more complicated example is the production of urea, a product of chemical reactions in the body (though the following equation does not show the whole process):

ammonia + carbon dioxide → urea + water
$$2NH_3 + CO_2 \rightarrow NH_2CONH_2 + H_2O$$

Note that some of these molecules have a number in the front of them, such as $2NH_3$. This 2 tells us that there are two of these molecules and that both take part in this reaction.

The direction of the arrow (→) is important too. It tells us that urea and water are made from ammonia and carbon dioxide. If you could see the reaction taking place, you would see the ammonia molecules and the carbon dioxide molecules gradually disappear as they combine into urea and water molecules. If the arrow had pointed in the opposite direction (←) it would mean that ammonia and carbon dioxide are made from urea and water. The urea and water molecules would have been replaced by ammonia and carbon dioxide molecules.

Reversible reactions

Sometimes you may see a reaction with arrows in both directions:

carbon dioxide + water $\rightleftarrows$ carbonic acid

This reaction also takes place in the body, when the carbon dioxide we either inhale or produce in our cells dissolves in the water in the body. The equation means that both reactions are taking place at the same time. Carbon dioxide is combining with water to make carbonic acid, but the acid is continually breaking apart into carbon dioxide and water. This is an example of a **reversible reaction**. If the two reactions occur at the same speed, the total reaction is said to be in equilibrium. In other words, the rate at which the reactants form products is equal to the rate at which the products reform the reactants. Equilibrium reactions have arrows in both directions of the same length.

Again, if you could see the numbers of molecules of reactants and products during equilibrium, you would find they remain fairly constant. As we go along, more will be said about equilibrium reactions. Some reversible reactions go in both directions at different rates; when this occurs it is shown by having the faster reaction with a longer arrow (e.g. $\rightleftarrows$).

Reversible reaction
One that can take place in both directions; that is, reactants form products at the same time as products reform the reactants. These reactions are indicated by double-headed arrows:

$\rightleftarrows$

Summary

1. The substances in the world are called chemicals.
2. Chemicals are made of atoms.
3. Neutral atoms have equal numbers of protons and electrons.
4. Atoms that have lost or gained electrons are called ions.
5. Two or more atoms of the same type join to form molecules of elements.
6. Two or more atoms of different types join to form molecules of compounds.
7. Mixtures contain separate, unjoined compounds and/or elements.

In the case of Laurel S. we have seen that two chemical compounds are important to her state of health: glucose and insulin. In the terms introduced so far in this chapter, both are molecules and both are compounds. The formula for glucose is $C_6H_{12}O_6$, which you now realise means one molecule, composed of six carbon atoms, 12 hydrogen atoms and six oxygen atoms. Insulin is a large, complex molecule with hundreds of atoms and a structure too complicated to depict easily.

4.2 BONDING
Ionic bonds

What holds molecules together? The quick answer to this question is electrostatic force. This force, remember, pulls together unlike charges (positive and negative) and repels like charges (positive/positive or negative/negative). In more detail, we can see this with an example with two ions:

positive sodium ion + negative chloride ion → salt
$Na^+ + Cl^- \rightarrow NaCl$

The positive and negative charges attract one another and as a result (in the case of NaCl) form a three-dimensional, cubic structure called a crystal lattice. Here are two more examples of this force acting:

K^+ + I^- →   KI   (+1   +   −1   =   0)
Ca^{2+} + $2Cl^-$ → $CaCl_2$ (+2   +   −2   =   0)

Ionic bond
Type of chemical bonding between ions, where positive ions are held together with negative ions by the electrostatic force of attraction.

These are examples of **ionic bonds**, which result from the attraction of oppositely charged ions. There are three things to notice about these bonds:

1. The charge on the molecule formed from the ions is the sum of the charges on the ions. This is shown in the brackets next to the examples.
2. When the number of atoms or ions is written in front of the symbol of the element, as in $2Cl^-$, we know how many of those chloride ions join with the calcium ion. That is,

1 calcium cation + 2 chloride anions → 1 molecule of calcium chloride
1 potassium cation + 1 iodide anion → 1 molecule of potassium iodide

3. There must be the same number of each atom on both sides of the arrow(s). If you start with 1 Ca and 2 Cl, you must end up with 1 Ca and 2 Cl. The molecule of $CaCl_2$ shows the 2 Cl by the subscript on the Cl. Subscripts, remember, show the numbers of that atom in the molecule. Figure 4.4 may help to make this clearer.

In all these diagrams of molecules, the bonds are represented by short straight lines. Atoms that are bonded together are shown connected with a bond line, or simply next to each other.

FIGURE 4.4

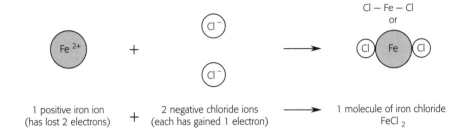

1 positive iron ion   2 negative chloride ions ⟶ 1 molecule of iron chloride
(has lost 2 electrons)   +   (each has gained 1 electron) $FeCl_2$

Covalent bonds

Ionic bonds are probably the easiest to understand, but there are other types of bonds holding molecules together. One of the most important of these is called the **covalent bond**. To understand how it works, we have to return to the structure of an atom.

The electrons circle the nucleus in orbits that chemists call energy levels. As we have mentioned, each energy level is at a specific distance from the nucleus. Every electron that is orbiting a nucleus has a specific amount of energy and those electrons with the same energy are found in an energy level technically called a shell. The shells closest to the nucleus have the lowest energy and those further away have greater energy. In addition, each shell can hold only a fixed number of electrons, as shown in Table 4.5.

As shown, shell 1, which has the lowest energy, can hold only two electrons. Once it is full, the third electron, if there is one, must go into the second shell. There are more electron shells than those shown in Table 4.5, as high as shell 7, but they are beyond the scope of this text.

We can show how the elements have their electrons arranged into shells by the use of simple diagrams like the ones in Figures 4.5(a) and (b).

Figure 4.5(a) shows the pattern for hydrogen and helium, the first two elements in the Periodic Table, and shows that, with helium, the first shell is full.

Covalent bond
Type of chemical bonding in which atoms share orbiting electrons. This sharing is done so that the atoms will have enough electrons to fill the outermost energy level of each atom.

TABLE 4.5 SOME ELECTRON SHELLS' CAPACITIES

Electron shell	Maximum electrons allowed
1	2
2	8
3	18
4	32

FIGURE 4.5(a) Electron arrangements in hydrogen and helium

FIGURE 4.5(b) Electron arrangements in lithium and neon

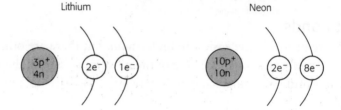

In Figure 4.5(b) the elements in the second row, lithium to neon, fill up the second shell; only lithium and neon are shown here.

A simpler way to show the arrangement of the electrons is given in Table 4.6, which shows the first 15 elements.

TABLE 4.6 ELECTRON ARRANGEMENTS OF THE FIRST 15 ELEMENTS

Element	Atomic number	Number of electrons in shell			
		1	2	3	4
Hydrogen	1	1			
Helium	2	2			(Shell 1 full)
Lithium	3	2	1		
Beryllium	4	2	2		
Boron	5	2	3		
Carbon	6	2	4		
Nitrogen	7	2	5		
Oxygen	8	2	6		
Fluorine	9	2	7		
Neon	10	2	8		(Shell 2 full)
Sodium	11	2	8	1	
Magnesium	12	2	8	2	
Aluminium	13	2	8	3	
Silicon	14	2	8	4	
Phosphorus	15	2	8	5	

And so it goes, filling shell by shell. Because some of the higher shells overlap, the shells do not always fill as neatly as the first 15 do, but that isn't important here.

Valence electrons and covalent bonds

The elements in the right-hand column of the Periodic Table, Group VIIIA (helium, neon, argon, krypton, etc.) are those with full shells. When chemists look at the properties of the different elements, they find that atoms with shells completely filled with the allowed numbers of electrons are very stable. That is, their atoms don't react with other atoms and therefore do not form compounds except under very unusual circumstances. For example:

Helium:	2 electrons	Energy level 1 full
Neon:	10 electrons	Energy levels 1 and 2 full
Argon:	18 electrons	Energy levels 1, 2 and 3 full

These three elements are never, in normal circumstances, found in compounds.

We can now return to our discussion of covalent bonds. Atoms whose shells are not full of electrons are therefore less stable and likely to form compounds by chemically reacting with other atoms. By reacting, it is possible for them to either gain enough extra electrons to fill completely their outermost shell, or to lose enough electrons, thereby emptying out their outermost shell. If they manage to do either, they will be more chemically stable.

Another way of filling that level is to share electrons with another atom. For example, hydrogen has only one electron in the first (and its outermost) energy level. If it had two, that level would be full. So it can either empty that level and become a positive ion (H^+) or it can fill that level by sharing some other atom's electron. A convenient way to do that is to join with another hydrogen so that they can both share the two electrons. This is called a single covalent bond (see Figure 4.6(a)) because a hydrogen atom is sharing one electron from another atom.

Another example is oxygen, which has eight electrons, six of which are in the

FIGURE 4.6(a) Single covalent bond

outer energy level, which can hold eight. With only two more, it would be as stable as neon, so it can form molecules by sharing two electrons with another atom or atoms. If it does, this is called a double covalent bond. This is the reason we usually find oxygen in the form of the molecule O_2. In Figure 4.6(b) it is represented by two bond lines.

FIGURE 4.6(b) Double covalent bond

A third example is carbon dioxide. Carbon has six electrons, four in the outer energy level, so it needs another four to be as stable as neon. It can share two with one oxygen atom and two with another, as Figure 4.6(c) shows.

FIGURE 4.6(c) Double covalent bond

The electrons in the outermost, unfilled shell that are available to be lost, gained or shared are referred to as valence electrons. They are extremely important in determining the chemical properties of their element. They occur in the valence shell, which is the outermost energy level of that atom. Looking at Table 4.6 again, we can see that the valence shell for lithium is shell 2 and that there is one valence electron in that shell. All the elements in that column of the Periodic Table, called 1A, have one valence electron. All the elements in the second column, 2A, have two valence electrons and so on, up to column 7A, with seven valence electrons. What this means is that we would expect the elements in column 1A to lose that valence electron, thus emptying the outermost shell, leaving them with only full shells. Elements in column 7A would be expected to gain one electron, thus filling their outermost shell.

Speaking more technically, if the outermost shells are filled by sharing valence electrons with other atoms, then this is referred to as a covalent bond. Covalent bonds can be formed between atoms of the same type, as in H_2, or different types, as in ammonia, NH_3. Atoms that are bonded through covalent bonds form a molecule of that compound. This sharing of electrons ties the molecule together. Covalent bonds, in fact, are often quite strong. Note too that in the examples that the molecule made by the atoms is electrically neutral.

One important type of covalent bond is called a **polar covalent bond**. In this case, the two atoms that share electrons are of different types and one atom gets a bigger share than the other. By 'bigger share' we mean that the shared electrons are pulled closer to that atom. The ability of an atom to attract the shared electrons is referred to as its electronegativity. The element with the greater electronegativity pulls the shared electron closer to itself. In the Periodic Table we find that those elements over to the right have the larger electronegativities, and an example of how the process occurs in water (H_2O) is given in Figure 4.7.

Partly because the oxygen atom has eight protons in its nucleus, and thus a charge of +8, it has a stronger pull on the shared electrons, each of which are −1 charge. These shared electrons thus tend to spend more time close to the oxygen atom rather than the hydrogen atoms. One thing this means is that the oxygen end of the molecule is slightly negatively charged and the hydrogen end is slightly positively charged. When this happens and the molecules end up with slightly charged ends, we refer to these as polar covalent bonds.

Polar covalent bonds
Bonds between pairs of atoms which result in the molecule having one end with a slight negative charge and the other end with a slight positive charge. In water, for example, the oxygen end of the molecule more strongly attracts the shared covalent electrons, giving that end the slight negative charge. Two such water molecules can hold together because of the electrical attraction between them; this is a **hydrogen bond**.

Hydrogen bonds

Thanks to the fact that the hydrogen end of a water molecule is slightly positive, it can be attracted towards the negatively charged oxygen end of another water molecule. It can then form a very weak bond to that molecule called a **hydrogen bond**. These bonds are usually too weak to hold a molecule together, but they can form weak links between nearby molecules. This is especially important in the structure of DNA (discussed in Chapter 11) and proteins. Speaking generally, then, a hydrogen bond results from the electrical attraction between

Hydrogen bond
A weak attraction between a hydrogen atom and an electronegative atom such as oxygen.

FIGURE 4.7 Polar covalent bond in H_2O.
The oxygen end of the molecule is negative compared with the hydrogen end.

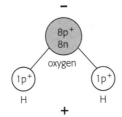

an electronegative atom (such as O, N, or F) and a hydrogen atom which is covalently bonded to another electronegative atom (usually O or N). This hydrogen bonding is characteristic of water (see Figure 4.8) and is essential to water's crucial role in the body.

4.3 ORGANIC COMPOUNDS IN THE BODY

Now we return to Laurel S. Part of her treatment consists of a carefully controlled diet. Her chart (Table 4.7) contains specific information recommended for diabetics by the Commonwealth Department of Health.

Now that we have some chemistry background, we can look more closely at how Laurel S.'s diet relates to her health. We'll take the items in the menu column and categorise them into distinct types.

Carbohydrates

Organic molecule
One containing carbon and which usually, but not necessarily, has been made by a living organism; for example, carbohydrates, lipids and proteins.

Carbohydrates are a group of molecules that contain carbon (C), hydrogen (H) and oxygen (O). They are **organic compounds**, which means they contain carbon and are most commonly made by living things. Examples of carbohydrates are starch and cellulose (plant fibre) and the various types of sugar, such as glucose and maltose. In Laurel S.'s diet, the carbohydrate items are cereal, bread, potato, biscuits and rice, though none of these is a pure carbohydrate.

FIGURE 4.8 Hydrogen bonds

Hydrogen bonds in water, formed as a result of electrical attractions between positively polarised hydrogen atoms and negatively polarised oxygen atoms. Each water molecule can bond to four neighbouring molecules.

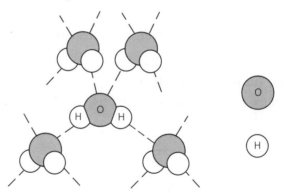

Source: *Essentials of General, Organic and Biological Chemistry*, McMurray, J., 1989, p. 125. © Reprinted by permission of Pearson Education Inc., Upper Saddle River, N.J.

TABLE 4.7 DIABETIC DIET: MAIN MEALS ONLY

Meal	Menu	Size of serve
Breakfast	Fruit juice	120 mL
	Cereal	1/2 cup
	Bread/toast	1 slice
	Hot dish — egg	1 large
	— lean meat	45 g
	— bacon/sausage	30 g
Lunch	Main dish/sandwich	
	— lean meat/fish	60 g
	— meat	40 g
	— cheese	30 g
	Potato/rice/spaghetti	75g
	Bread	1 slice
	Salad/vegetables	as desired
	Fresh/stewed fruit	120 g/1 piece
Main meal	Main dish	
	— lean meat/fish	60 g
	— meat	40 g
	Potato/rice/bread	75 g/1 slice
	Vegetables	as desired
	Fresh/stewed fruit	120 g/1 piece
	Milk puddings (from daily milk allowance)	300 mL

Source: Commonwealth Dept. of Health (1980) *Hospital Diet Manual*, AGPS, Canberra.

Notice the absence of cane sugar (sucrose) in these food items; there are no cakes, for example. The sources of sweetness here are the fresh or stewed fruits, which contain either fruit sugar (fructose) or glucose.

Some of the important carbohydrates in nursing

Carbohydrates are named either from the number of carbon atoms in them, or from their old Greek and Latin names. For example, carbohydrates with five carbon atoms per molecular unit are called pentoses, from the Greek word *pente*, for five. $C_5H_{10}O_5$ is therefore one type of pentose. If there are six carbons, we have hexoses and so on. Notice that they all end in the suffix -ose (meaning sweet). The ones mentioned above are also referred to as monosaccharides (from mono-, for one and saccharide, sugar). They cannot be broken down to a simpler type of sugar.

Sometimes the source of the carbohydrate is used to name them. In the human body, the most important monosaccharides are glucose (from the Greek word for sweet), fructose (found in fruit and honey) and galactose. Glucose, fructose and galactose all have the same chemical formulae: $C_6H_{12}O_6$. They differ only in the way the atoms are arranged in the molecule. By the way, fructose is the sweetest of the carbohydrates, twice as sweet as sucrose (table sugar), making it popular with those on a diet, who want the same sweetness with less calories. Figure 4.9 shows the structure of the monosaccharides named above.

Disaccharides (from di-, meaning 'two') are sugars made up of two monosaccharides, and have the general formula $C_{12}H_{22}O_{11}$. Examples of disaccharides are maltose, lactose and sucrose. The structure of sucrose is shown in Figure 4.10.

As we've said, Laurel S.'s diet contains no sucrose, but it may come as a surprise to you that sucrose, which comes from sugar cane and sugar beet, is produced in greater quantities around the world than any other organic compound.

Polysaccharides (from poly-, meaning 'many') are long chains of monosaccharides, strung like beads on a necklace. Three well-known polysaccharides are starch, cellulose and glycogen. They are all chains of glucose, differing mainly in where the chain branches. Some starches may have from 250 to 4000 glucose units strung together. This is shown in abbreviated form in Figure 4.11.

FIGURE 4.9 Structure of three common monosaccharides

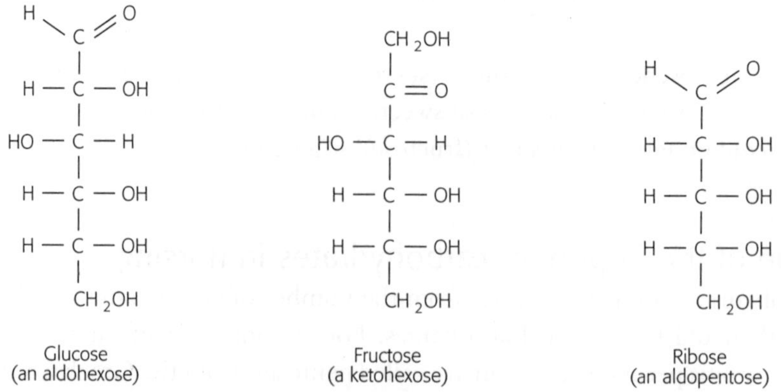

Glucose
(an aldohexose)

Fructose
(a ketohexose)

Ribose
(an aldopentose)

Source: *Essentials of General, Organic and Biological Chemistry*, McMurray, J., 1989, p. 275. © Reprinted by permission of Pearson Education Inc., Upper Saddle River, N.J.

FIGURE 4.10 The structure of sucrose

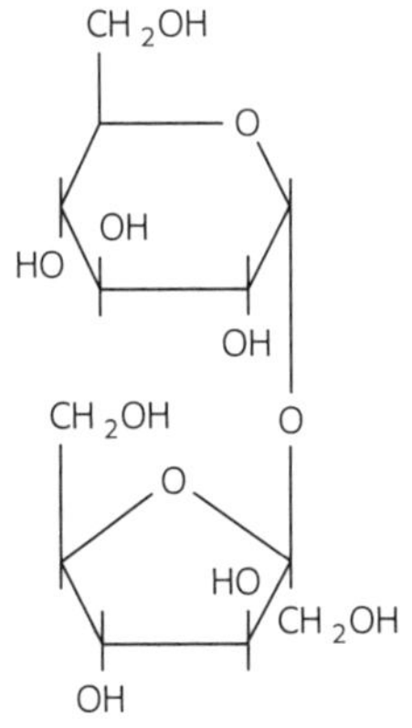

FIGURE 4.11 Simplified polysaccharide structures

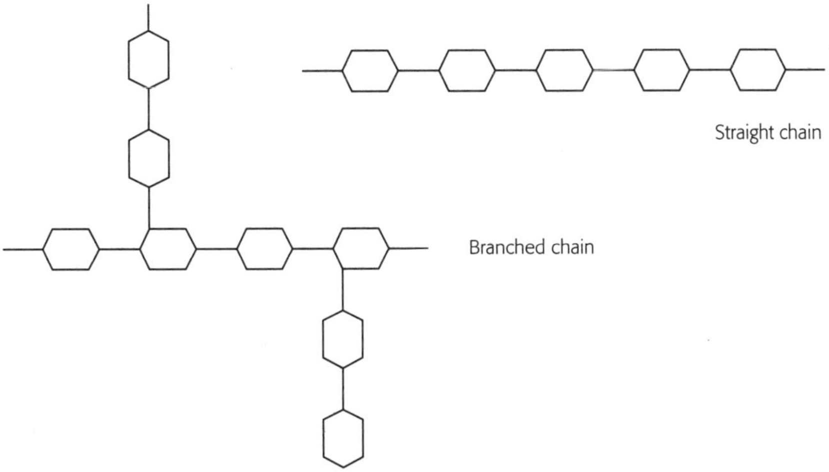

Straight chain

Branched chain

Why carbohydrates are important

Carbohydrates are the main energy source of the body. In the case of Laurel S., she needs to reduce weight. If she takes in more carbohydrate than is needed for energy, the rest can be converted to fat and stored around her body. Therefore her intake of carbohydrate must be carefully controlled. How carbohydrate is converted to energy or fat inside the body is discussed in the next chapter. However, diabetes is a complex condition that involves, among other things, the control of glucose levels in the body, and more than simple dietary control is necessary for Laurel S.

Glucose requires no digestion and can be given intravenously (when it's called dextrose) to clients who cannot take food by mouth. It is often found in the urine of clients who have diabetes mellitus (or 'sweet' diabetes). If this occurs, the client is said to suffer from glycosuria.

As mentioned, fructose occurs naturally in fruit and honey. Some newborn infants have fructosaemia, a condition that prevents their bodies from using fructose. The symptoms are vomiting, severe malnutrition and hypoglycaemia (or low levels of glucose in the blood). The standard treatment is dietary; the infant is put on a diet with little or no fructose in it.

Galactose is formed from the lactose in breast milk and is also found in sugar beets. Again, some babies cannot tolerate this form of hexose and suffer from galactosaemia. This is an inherited disease and the child has to be fed on other sources of sugar.

Clinical note: Normal blood glucose levels range from around 50–120 mg/dL, rising and falling with the intake in the diet. Hypoglycaemia refers to a clinical condition where the blood sugar levels range consistently between 40 mg/dL and 100 mg/dL, despite adequate glucose in the diet. Hyperglycaemia refers to blood glucose levels consistently ranging from 120 mg/dL to 350 mg/dL.

Where carbohydrates come from
Carbohydrates are made by plants during photosynthesis:

$$\text{carbon dioxide} + \text{water} \xrightarrow{\text{sunlight}} \text{glucose} + \text{oxygen}$$
$$6CO_2 + 6H_2O \rightarrow C_6H_{12}O_6 + 6O_2$$

This glucose can then be joined with others to make more complex carbohydrates, like sucrose or starch.

Important carbohydrate chemical reactions
There are basically two important reactions. The first of these takes simple sugars (monosaccharides) and joins them together to make a longer, more complex molecule. For example, two monosaccharides can join to make a disaccharide:

$$\text{glucose} + \text{glucose} \rightarrow \text{maltose} + \text{water}$$

In symbols, this is:

$$C_6H_{12}O_6 + C_6H_{12}O_6 \rightarrow C_{12}H_{22}O_{11} + H_2O$$

Because this reaction removes a water molecule from the two glucose molecules, it is referred to as **dehydration** synthesis.

The second reaction is the reverse of the one above. A complex carbohydrate is broken down into simpler molecules:

$$\text{maltose} + \text{water} \rightarrow 2 \text{ glucose molecules}$$
$$C_{12}H_{22}O_{11} + H_2O \rightarrow 2C_6H_{12}O_6$$

This reaction is called **hydrolysis** (*hydro* means 'water' and *lysis* means 'breaking apart', or 'dissolution'). Hydrolysis is the reaction of a compound with water. The above reaction normally occurs only very slowly unless an acid or catalyst is present. Other examples are:

$$\text{lactose} + H_2O \rightarrow \text{glucose} + \text{galactose}$$
$$\text{sucrose} + H_2O \rightarrow \text{glucose} + \text{fructose}$$

Speaking technically, we would say that sucrose has been hydrolysed into two monosaccharides. Similarly, it is correct to say that a polysaccharide such as glycogen can be hydrolysed in the cells to give glucose for energy. More will be said about carbohydrates and their energy role in the body in Chapter 5.

Lipids

The most common lipids are known as simple lipids or fats, though the complete class of lipids includes such compounds as oils, waxes, steroids and more specific groups such as prostaglandins, phospholipids and glycolipids. Lipids too are organic compounds. Common examples are butter, peanut oil and ear wax, though none of these is a pure lipid, but a mixture. They form about 12% of the human body. In Laurel S.'s diet, the foods that contain lipids are the cheese, milk pudding and the butter or margarine on her toast. Among their most important characteristics is the fact that they are insoluble in water, but will dissolve in another lipid. Since lipids are a major part of the cell membrane, this is very important, assisting many water-insoluble substances to enter the cell, or adhere to its surface.

Chemical reaction
The term used for the joining together of atoms, ions or molecules and for the splitting apart of molecules into either atoms, ions or smaller molecules.

Hydrolysis reaction
One in which water molecules take part; when added to a reactant, water breaks it down into smaller products.

Dehydration reaction
The opposite of an hydrolysis reaction; in this case, one of the products is water, the elements of which have been removed from one or more of the reactants.

The structure of a lipid

Lipids, like carbohydrates, contain mostly carbon, hydrogen and oxygen. They may also contain small amounts of nitrogen (N), or phosphorus (P), as in the phospholipids. They are generally much bigger molecules than either monosaccharides or disaccharides and usually form long chains or rings of many atoms. Since there are so many different lipids in the body, it is best to concentrate on the most important types of lipids presented in Figure 4.12.

Lipids are classified in terms of their chemical structure and reactions. In general terms, a lipid is made from one or more fatty acids and an alcohol. That is:

$$\text{lipid} + \text{water} \rightarrow \text{fatty acid(s)} + \text{an alcohol}$$

This is another example of an hydrolysis reaction and the reaction normally goes slowly unless an alkali or catalyst is present.

What is a fatty acid? It's the long chain part of the lipid molecule. For example, one of the fats in peanut oil contains a fatty acid called stearic acid, $C_{17}H_{35}COOH$. The 17 carbon atoms are arranged in a long chain. They are held together by covalent bonds. (A molecular model kit would be very useful; you could construct a stearic acid molecule to see both the chain and the interesting three-dimensional shape of this compound.) The COOH group of atoms is written together to show that, in the molecule, they form a separate group. This particular combination of atoms is referred to as a carboxyl group. Such specific combinations are referred to in general terms as functional groups, because they are essential in giving the molecule its special characteristics, or

TABLE 4.8 CLASSES OF LIPID MOLECULES

Lipids	Composition
Waxes	Fatty acid and long-chain alcohol
Fats and oils (triglycerides)	Fatty acids and glycerol
Phospholipids	Fatty acids, glycerol, phosphate, amino alcohol
Sphingolipids	Fatty acids, sphingosine, phosphate, amino alcohol
Glycolipids	Fatty acids, glycerol or sphingosine, one or more monosaccharides
Steroids	A fused structure of three cyclohexanes and a cyclopentane

Source: *Chemistry*, 7th Ed, Timberlake, Karen C., 1999, p. 559. © 2001 by Benjamin Cummings. Reprinted by permission of Pearson Education Inc.

FIGURE 4.12 Some important types of lipids

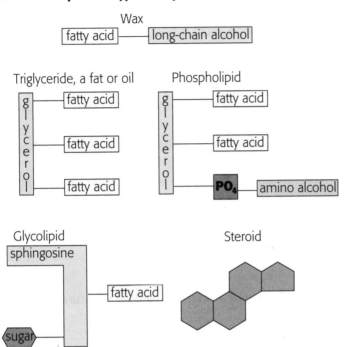

Source: *Chemistry*, 7th Ed, Timberlake, Karen C., 1999, p. 559. © 2001 by Benjamin Cummings. Reprinted by permission of Pearson Education Inc.

functions. More is said about functional groups later in this chapter. Figure 4.13 shows the structure of a stearic acid molecule, including the carboxyl group.

The carboxyl group can give the properties of an acid to the molecule it is attached to, and therefore any molecule that contains a carboxyl group is said to be a type of carboxylic acid. (Acids are discussed in detail in Chapter 7.) Thus, for lipids, we refer to this molecule as a fatty acid and stearic acid is one example of a fatty acid.

If each of the 17 carbon atoms in the stearic acid chain forms only single covalent bonds to two other carbons, one on each side, like this: C–C–C–C–

FIGURE 4.13 Structure of a stearic acid molecule $CH_3(CH_2)_{16}COOH$

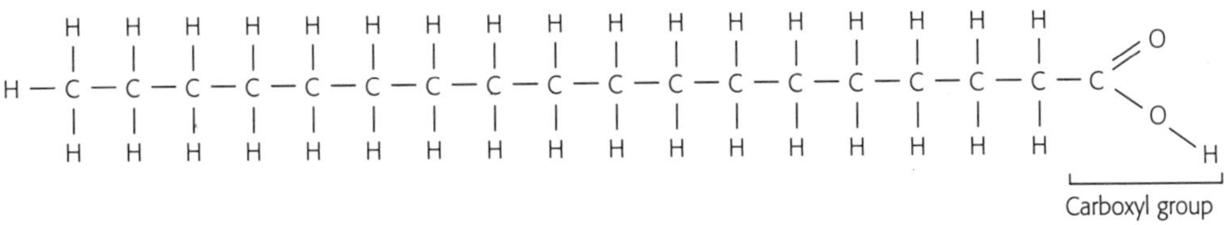

Carboxyl group

Saturated and unsaturated fatty acids
Saturated fatty acids have no double bonds between the carbon atoms in the chain; thus they are saturated with hydrogen atoms. Unsaturated fatty acids have fewer hydrogen atoms, because some of the bonds are in the form of double bonds between carbon atoms.

etc., then the fatty acid is said to be saturated. **Saturated fatty acids**, then, have only single carbon–carbon bonds, with hydrogen atoms taking up the other carbon bonds. **Unsaturated fatty acids** have one or more double covalent bonds between the carbons, like this: C=C–C=C–C=C etc. Stearic acid is a saturated fatty acid, but oleic acid, found in olive oil, is an unsaturated fatty acid. Its formula is $C_{17}H_{33}COOH$. Note that it has two less hydrogens than stearic acid ($C_{17}H_{35}COOH$), because two of the carbons are using double covalent bonds. Fats and oils can contain three different types of fatty acid molecules: saturated, unsaturated, or some combination of these; examples are shown in Figure 4.14. Laurel S.'s diet contains unsaturated fats.

We can make most of the fatty acids we need from the carbohydrates in our diet. However, this is not possible if the fatty acids need more than one double bond; these we must get in the diet. They are therefore referred to as essential fatty acids.

What about the alcohol part? Alcohol is really a general name for a group of

FIGURE 4.14 The structure of a saturated and an unsaturated fatty acid

$$CH_3CH_2CH_2CH_2CH_2CH_2CH_2CH_2CH_2CH_2CH_2CH_2CH_2CH_2C-OH$$

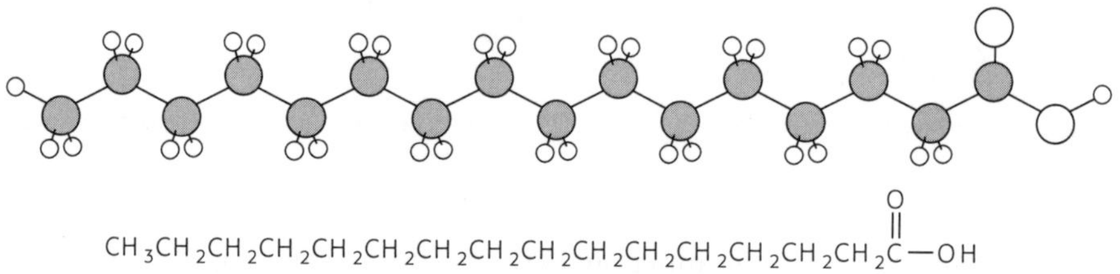

Palmitic acid—a saturated fatty acid

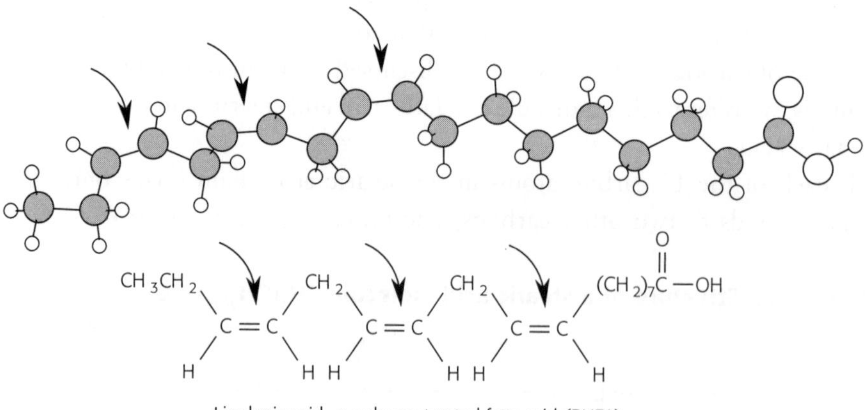

$$CH_3CH_2 \diagdown C=C \diagup CH_2 \diagdown C=C \diagup CH_2 \diagdown C=C \diagup (CH_2)_7C-OH$$

Linolenic acid—a polyunsaturated fatty acid (PUFA)

Source: *Essentials of General, Organic and Biological Chemistry*, McMurray, J., 1989, p. 300. © Reprinted by permission of Pearson Education Inc., Upper Saddle River, N.J.

molecules with certain chemical similarities. All alcohols contain only carbon and hydrogen, but have at least one –OH functional group replacing one of the hydrogens. Examples are methanol: CH_3–OH; and glycerol (commonly known as glycerine): $CH_2OH.CHOH.CH_2OH$.

The most common types of lipids are fats and oils. These are also known as triglycerides, because they have three fatty acids joined to one glycerol, as shown in Figure 4.15.

A triglyceride therefore breaks down (hydrolyses) to fatty acid(s), glycerol, and in some cases other compounds, such as a carbohydrate.

Why lipids are important

Laurel S. is having problems with high blood pressure due to the build-up of cholesterol in her arteries, which narrows them. Cholesterol is a lipid with an interesting structure, shown in Figure 4.16.

See how the carbons form hexagons and a pentagon. (Cholesterol is another good molecule to make with a molecular model kit.) Cholesterol is not all bad, however. It is a necessary component of all cell membranes and helps the small intestine's absorption of fatty acids.

Other important lipids are steroids, prostaglandins and phospholipids.

FIGURE 4.15 General structure of fats and oils (triglycerides)

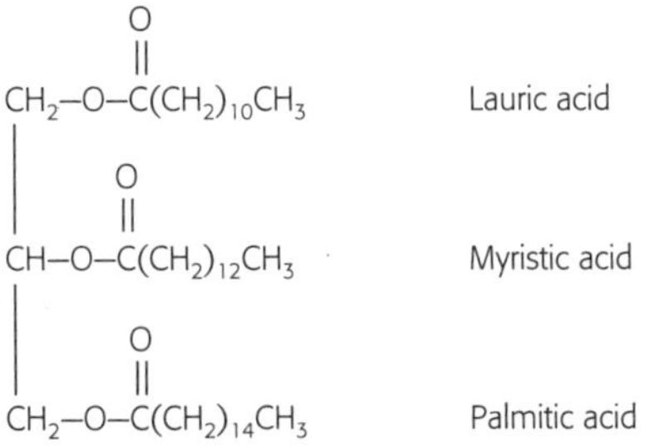

FIGURE 4.16 Cholesterol, an unsaturated steroidal alcohol

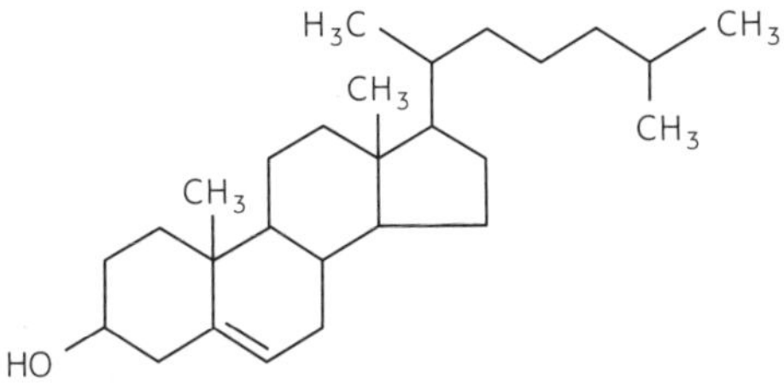

Source: *Essentials of General, Organic and Biological Chemistry*, McMurray, J., 1989, p. 308. © Reprinted by permission of Pearson Education Inc., Upper Saddle River, N.J.

Steroids

These are large lipids and cholesterol is an example. Male and female sex hormones are also steroids. Their use (and abuse) in sport is becoming an important health-related issue. Vitamin D is another well-known steroid.

Prostaglandins

These are hormones that control the behaviour of neighbouring cells. They have a very wide range of influences in the body, especially over maintaining homeostasis. For example, prostaglandins are involved in the sensation of pain, control of body temperature (as explained in Chapter 2) and the release of insulin.

Phospholipids

All cell membranes contain phospholipids. They literally hold the contents of the cell inside and help to control the movement of chemicals into and out of the cell. Since they are lipids, they therefore tend to make the cell 'waterproof' too. Figure 4.17 shows the structure of a typical phospholipid.

Lipids certainly play an important role in the body. At this point, it is useful to remember the following:

1. Lipids are long-chain or ring-shaped molecules, mainly of carbon, hydrogen and oxygen.
2. They are generally insoluble in water.
3. They help form the cell membrane.
4. They are the basis of many hormones and steroids.
5. They are a source of energy.
6. They insulate the body and also have a protective, cushioning role.
7. Triglycerides are formed from fatty acids and glycerol, using dehydration synthesis reactions.

Lipoproteins

Because lipids do not dissolve in water, they are not easily transported by the bloodstream around the body. This problem is solved by having the lipids attach themselves to proteins and phospholipids, forming soluble complexes called *lipoproteins*. These come in four types, which differ in their density and type of lipid. You may have heard of these in other contexts.

FIGURE 4.17 The structure of a typical phospholipid

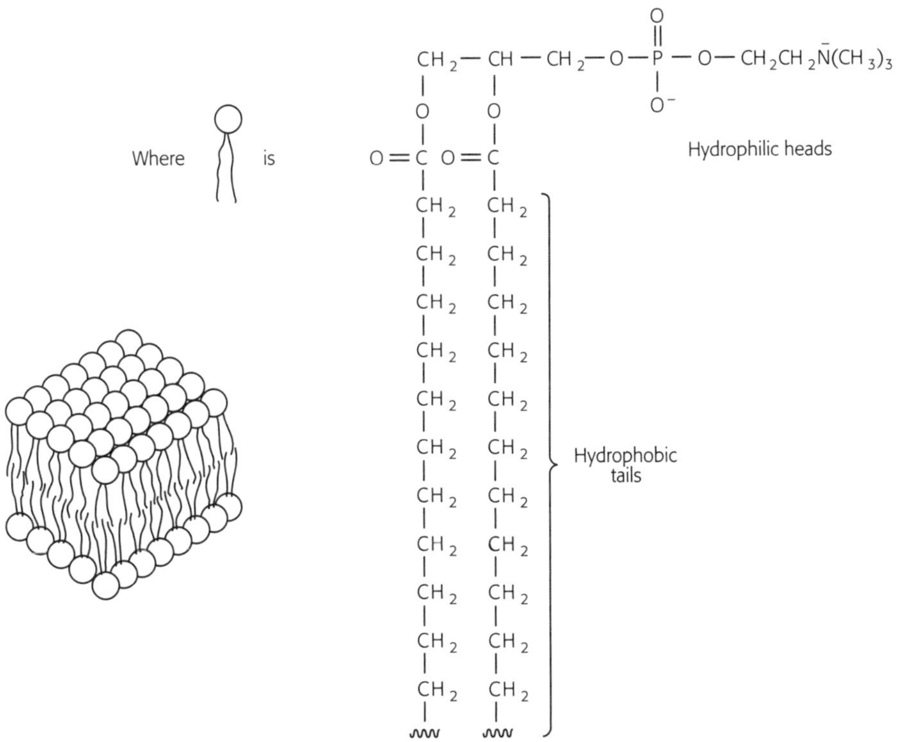

Source: *Essentials of General, Organic and Biological Chemistry*, McMurray, J., 1989, p. 306. © Reprinted by permission of Pearson Education Inc., Upper Saddle River, N.J.

There are *very low density lipoproteins* (VLDLs) and the *chylomicrons* which carry triglycerides between the intestines or liver and the body cells.

The *low density lipoproteins* (LDLs) carry cholesterol to the tissues.

The *high density lipoproteins* (HDLs) carry cholesterol from the tissues to the liver for the production of bile salts, which are then excreted.

It is common these days to test the blood for the levels of LDLs and HDLs as an indicator of risk of atherosclerosis. High levels of LDLs are associated with high cholesterol levels, which can deposit in cells or arteries, reducing their diameter. High levels of HDLs are considered a good sign, as it indicates that cholesterol is being removed from the cells, lowering the risk of deposition in the arteries.

Proteins

Proteins are the third of our main organic molecule groups. The items in Laurel S.'s diet that contain protein are the meats and fish. The word 'protein' means 'primary', which describes well their importance in animal life. Of course, we need carbohydrates and lipids as well, but proteins are so central to the life of the cell and to whole organisms that many scientists of the past thought of them as containing life itself. There are so many types of proteins that it may be best to start with a description of their structure.

Protein structure

Figure 4.18 shows one of the units of a protein, an amino acid called alanine.

All proteins are made by joining together amino acids. There are about 20 different amino acids in nature and they combine to form all of the roughly 100 000 kinds of protein found in the human body. Amino acids all contain nitrogen, usually as a NH_2 group, called an amine group. They all contain a COOH group, already described as a carboxyl group, but also called a carboxylic acid group. This is why they are called amino acids. Figure 4.19 shows both these functional groups.

When amino acids join together, they form larger molecules. For example, two amino acids can join to form a molecule known as a dipeptide.

glycine + alanine → a dipeptide + water

In this case, the name of the dipeptide formed is glycylalanine, which is often abbreviated as (Gly-Ala). The bond is referred to as a peptide bond. As you can imagine, we can then form tripeptides, tetrapeptides, etc. Really long chains of amino acids, up to 50 units long, are called polypeptides. The name 'protein' is

FIGURE 4.18 Alanine, an amino acid

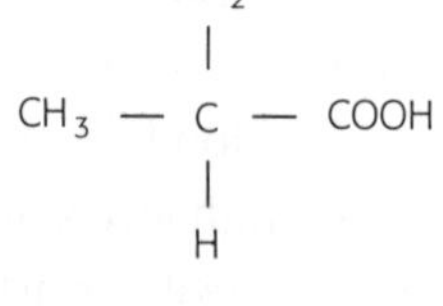

$$CH_3 - \underset{\underset{H}{|}}{\overset{\overset{NH_2}{|}}{C}} - COOH$$

FIGURE 4.19 Aspartane, an amino acid, showing the NH_2 amine group and the COOH carboxyl group

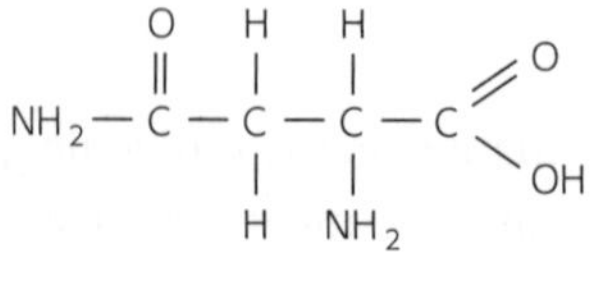

$$NH_2 - \overset{\overset{O}{\|}}{C} - \underset{\underset{H}{|}}{\overset{\overset{H}{|}}{C}} - \underset{\underset{NH_2}{|}}{\overset{\overset{H}{|}}{C}} - C \overset{\diagup O}{\diagdown OH}$$

usually reserved for chains greater than 50 units long. Some proteins are very long indeed, up to 2000 amino acids in length.

The structure of protein molecules can be very complicated. The sequence of the amino acids that are linked together by peptide bonds is called the primary structure of that protein. These are usually written by using the abbreviations of the amino acids; for example, Gly-Ile-Val-Glu-Gln- etc.

The polypeptides may then twist into shapes such as a coiled spring, where the coils are held in place by hydrogen bonds between them. This secondary structure results in complex three-dimensional shapes. And finally, some of these three-dimensional polypeptides can form weak bonds with other polypeptides to form the complete protein with a tertiary and sometimes quaternary structure. The important thing about all this is that shape is responsible for what the protein does. This is explained more fully in the section on enzymes in the next chapter, but the haemoglobin molecules in Figure 4.20 show the complexity of the three-dimensional structure.

Some important proteins that you will meet frequently in your nursing are included in Table 4.9.

Laurel S. is a diabetic and the amount of the particular protein called insulin in her bloodstream is important. However, as Table 4.9 suggests, every aspect of her normal body function involves protein, which is essential for support,

FIGURE 4.20 The structure of haemoglobin.

Haemoglobin consists of four globular protein subunits. Each subunit contains a single molecule of heme, a porphyrin ring surrounding a single atom of iron.

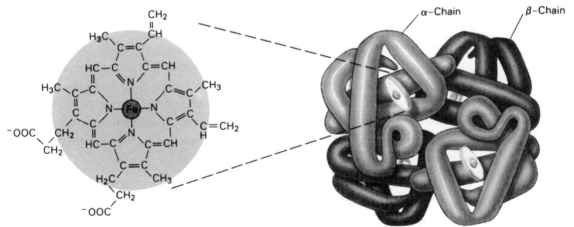

Haemoglobin molecule

TABLE 4.9 CLASSIFICATION OF SOME PROTEINS AND THEIR FUNCTIONS

Type of protein	Function in the body	Examples
Structural	Provide structural components	*Collagen* is in tendons and cartilage. *Keratin* is in hair, skin, wool, and nails.
Contractile	Movement of muscles	*Myosin* and *actin* contract muscle fibres.
Transport	Carry essential substances throughout the body	*Haemoglobin* transports oxygen. *Lipoproteins* transport lipids.
Storage	Store nutrients	*Casein* stores protein in milk. *Ferritin* stores iron in the spleen and liver.
Hormone	Regulate body metabolism and nervous system	*Insulin* regulates blood glucose level. *Growth hormone* regulates body growth.
Enzyme	Catalyse biochemical reactions in the cells	*Sucrase* catalyses the hydrolysis of sucrose. *Trypsin* catalyses the hydrolysis of proteins.
Protection	Recognise and destroy foreign substances	*Immunoglobulins* stimulate immune response.

Source: *Chemistry*, 7th Ed, Timberlake, Karen C., 1999, p. 595. © 2001 by Benjamin Cummings. Reprinted by permission of Pearson Education Inc.

movement, transport of important molecules, digestion, protection and reproduction.

To remain healthy, Laurel S., like the rest of us, must have a supply of all 20 amino acids. It is interesting that many plants can make all 20 from the raw materials (the C, H, O, N, etc. in their environment), but we can make only 10. The remainder must be given to us, readymade, in our diet. The ones we cannot make are called the essential amino acids, though of course all of them are essential for health. The ones we can make, the so-called non-essential amino acids, get their carbon atoms from molecules such as glucose and their nitrogen from ammonia (NH_3) or other amino acids.

Protein hydrolysis and denaturation

The catabolism of proteins is an important aspect of their biochemistry for two main reasons. One, the proteins that enter the body in the diet have to be broken down by digestive activities, and this is best understood by referring to the chemical reactions involved. And two, there are many causes of the destruction of proteins in the body, some of which are of clinical importance for understanding illness or disease.

Proteins, or more specifically the peptides of which they are made, can be broken apart by adding water to the bonds. This is another example of a hydrolysis reaction, which was mentioned earlier. This reaction needs an enzyme or an acid to be present at the same time. Using our example of (Gly-Ala) above, the hydrolysis reaction would look like this:

$$\text{glycylalanine} + H_2O \xrightarrow{H^+ \text{ or acid}} \text{glycine} + \text{alanine}$$

Denaturation of a protein refers to a disruption to the structure of a protein molecule. This disruption can affect the secondary, tertiary or quaternary structure. The protein then loses its original shape and, as we have seen, that will alter its function.

Substances capable of denaturing proteins include acids and bases, heat, other organic molecules, heavy metal ions, and physical shaking. A quick glance at these can be of interest to us here.

Heat denatures protein by breaking the hydrogen bonds. The most familiar example of this, of course, is cooking, where we alter the structure of the protein (making it easier to digest) but not its nutritional value. The use of very high temperatures can also sterilise surgical instruments, by destroying the protein structure of any micro-organisms present.

Acids and bases also break hydrogen bonds, but ionic bonds are also affected. For example, some burn ointments contain tannic acid, which causes proteins to clump together, forming a protective layer over the burn and preventing fluid loss.

Organic molecules include such things as alcohols, which are used to disinfect. In this case, they form their own hydrogen bonds with the protein, altering its shape and function.

Some metals such as iron (Fe^{2+}), lead (Pb^{2+}) and mercury (Hg^{2+}) form their own ionic bonds with the protein.

And shaking or agitation, such as when we whip cream or beat an egg, stretches the protein until the bonds break.

Nucleic acids

Another important group of organic molecules are the nucleic acids. You may have heard of them by their common abbreviations, DNA and RNA. DNA stands for deoxyribonucleic acid and RNA stands for ribonucleic acid. These large molecules contain the genetic information of each cell. Coded in their structure is all the information needed to determine whether the cell is that of a plant or an animal; whether that animal is fish or human; whether that human is male or female; whether it is destined to act as liver cell or muscle cell. The details of the structure and function of DNA and RNA are given in Chapter 11. Here, we will only briefly describe their components.

Like carbohydrates and proteins, nucleic acids are made from chains of smaller molecules. Nucleic acids are polynucleotides. Each one of these nucleotides consists of three parts. One is a pentose form of sugar, called an aldopentose, shown in Figure 4.21(a).

In DNA, the aldopentose is missing one of its oxygen atoms, which is why we refer to it as deoxyribonucleic acid.

The second part is one of five possible amine groups, where the word 'amine' means they contain nitrogen. One of these, thymine, is shown in Figure 4.21(b).

In DNA the possible amine groups are adenine, guanine, cytosine and thymine; in RNA, thymine is replaced by uracil. The third nucleotide component is a phosphoric acid molecule, H_3PO_4; see Figure 4.21(c). If we put these three together, we have a nucleotide, and when these are strung together,

FIGURE 4.21(a) An aldopentose

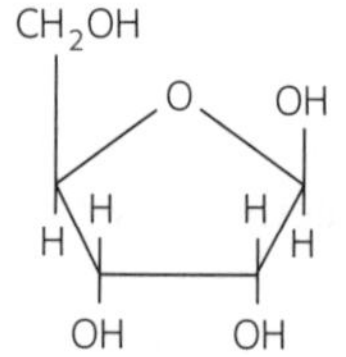

FIGURE 4.21(b) Thymine

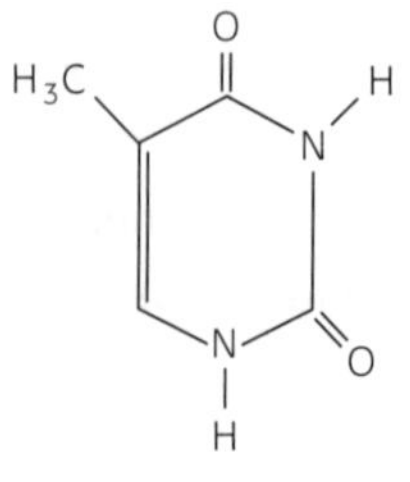

FIGURE 4.21(c) A phosphoric acid molecule

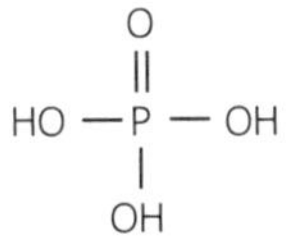

$$\begin{array}{c} O \\ \| \\ HO - P - OH \\ | \\ OH \end{array}$$

we then have either a RNA molecule (if we use the ribose sugar) or a DNA molecule (if we use the deoxyribose sugar). As mentioned above, the details of this structure are given in Chapter 11.

4.4 FUNCTIONAL GROUPS

As we have seen, many organic molecules have structures and reactions that are closely related. For example, the collection of molecules known as alcohols has certain similarities in structure. They all contain –OH as a group. In fact, the presence of an –OH group is enough to allow the chemist to recognise the molecule as one type of alcohol. When we discussed the fatty acids earlier in this chapter, we mentioned another such group called a carboxyl group. A carboxyl group has –COOH as its combination of atoms.

The general term given to such combinations is *functional group*, which can be defined as an atom or group of atoms that has a strong influence over the chemical behaviour of an organic compound. Why does it have such an influence? It is because most organic molecules have one particular place or section on the molecule where reactions typically take place. These sections are, as you would expect, where the functional group is found. These functional groups are useful to us because they help us to get a quick idea of the chemical behaviour of a molecule we might never have come across before. Being told that it contains the alcohol functional group, for example, tells us that we can expect it to behave in ways similar to the alcohols we are already familiar with, such as those used for disinfecting.

Functional groups that are commonly encountered in nursing applications are:

- alcohol (OH)
- amino group (NH_2)
- carboxyl (Carboxylic acid) (COOH)
- hydroxyl (OH)
- ketone (CO)

4.5 CHEMICAL HOMEOSTASIS

We have spent a lot of time introducing chemistry for nurses for some very important reasons. First, it's important for describing the client's condition. To say that Laurel S. has glycosuria, for example, immediately tells you something about her state of health. Second, chemical knowledge helps us understand how the body functions. We know more about lipid digestion, for example, when hydrolysis and hydrogen bonds are understood. Third, of course, all drugs are chemicals. If you are told, for example, that aspirin relieves pain because it hinders the production of prostaglandins (which are involved in pain), you now have some idea of what that means.

All of the above are separate parts of the same process, homeostasis. The body must control its internal environment and that includes the chemical part too. The types of molecules present and the amount of each are carefully controlled in healthy states. Many diseases and states of ill-health such as obesity or anorexia nervosa are dangerous simply because they threaten homeostasis. In the following chapters we will see over and over again how crucial this is. But as an example of this process, recall the homeostatic regulation of insulin outlined in Chapter 2. Since Laurel S. is diabetic, this is particularly relevant to her case.

After a meal, the level of glucose and amino acids in the blood rises. This stimulates the beta cells in the pancreas to manufacture insulin and release it into the bloodstream. The insulin affects skeletal muscle, the liver, fatty tissue and other tissues, causing them to increase their intake and use of glucose and amino acids. This is done by changing the way the cell membranes allow substances in and out. As the levels of glucose and amino acids in the blood decrease, the rate of insulin production also goes down.

This is, of course, another negative feedback loop, part of a much larger, complicated process of controlling the amount of nutrients such as glucose in the blood. Without sufficient insulin, this process cannot work efficiently. The treatment, whether by diet or insulin therapy, is clearly designed to correct a chemical imbalance.

Questions

Level 1

1. Distinguish between a compound and a mixture. Which of the following are mixtures: sea water, milk, a cylinder full of carbon dioxide, an apple?
2. Describe the structure of an atom, using the terms 'nucleus', 'electron', 'proton' and 'neutron'. How does an atom differ from a molecule? From an ion?
3. Use the Periodic Table to find the number of protons and electrons in the following atoms: hydrogen, carbon, oxygen, sulphur, chlorine.
4. What elements and how many atoms of each, are present in the following compounds:
   $KClO_3$ $FeSO_4$ $K_2Cr_2O_7$ H_2SO_4 $2Cu(NO_3)_2$
5. What is a double covalent bond?
6. What are some of the important types of lipids? Name some of their functions in the body.
7. What is characteristic about the chemical composition of proteins? Why are amino acids given that name?
8. What is the difference between an essential amino acid and a non-essential amino acid?

Level 2

9. What is the difference between ionic bonds and covalent bonds?
10. Why does a covalent bond share electrons?
11. How can hydrogen bonds help to explain why water molecules are weakly attracted to one another?
12. How would you distinguish between a carbohydrate and a lipid? What is the difference between:
    (a) a monosaccharide and a disaccharide?
    (b) a saturated fat and an unsaturated fat?
13. Distinguish between an hydrolysis reaction and a dehydration synthesis reaction. Give an example of each.
14. Describe the structure of a typical nucleotide.

Level 3

15. Discuss the concept of chemical homeostasis. What do you think is meant by the statement: 'Life is simply the sum of the chemical reactions taking place within the organism'?

16. Explain how Laurel S.'s dietary treatment is helping to deal with her diabetes and her obesity. Be specific in your discussion, referring to insulin, glucose and the organic molecules in her diet.

CALCULATION PROBLEM RELATED TO THE CASE STUDY

Laurel S. is ordered 36 units of 'Actfast' insulin each morning and 18 units each night. What are the two volumes you would give if the 10 mL vial contains 100 U/mL?

Answer:

Morning: Injection volume = 36 U/100 U × 1 mL = 0.36 mL of Actfast.

Evening: Injection volume = 18 U/100 U × 1 mL = 0.18 mL of Actfast.

THE GENERAL WARD AND CLIENT CARE: PART 2

METABOLISM, ENZYMES AND CHEMICAL ENERGY

Chapter outline

5.1 **Metabolism.** Your objective: to review and understand some simple chemical reactions associated with metabolism, using the example of digestion.

5.2 **Enzymes.** Your objective: to understand the structural features of enzymes and their role in metabolism.

5.3 **Energy production in the body.** Your objectives: to understand the process of cellular respiration; to know the structure and function of ATP; and to understand the nature of oxidation–reduction reactions.

INTRODUCTION

The introductory chemistry in the last chapter can be put to further good use as this chapter looks at more applications of chemical science to client care. It will help you to understand some of the problems associated with chemical imbalance in the body. In particular, we'll talk about how the enzymes control the rate of chemical reactions that occur within the body, and how the body obtains energy from the diet. Along the way we will discuss briefly some conditions that result from a breakdown in chemical homeostasis.

Case Study: Ms Rachel L.

Rachel L. has only recently come up to the general ward from Accident and Emergency. She was admitted comatose, suffering from alcohol poisoning. Only 19 years old, she has a history of alcohol-related problems, and suffers from hepatitis C. On admission she weighs 50 kg, is suffering some malnutrition and has hypoglycaemia. She is being given an intravenous dextrose solution. Her chart contains the following information.

Time: 0900. Intravenous therapy begun. 20 gauge angiocatheter inserted into left vein of forearm. 1000 mL D5W mEq KCl added. Solution time taped to run for 8 hours at 125 mL/hr. Drip rate 42 drops/min. NPO. Client voicing no complaints regarding intravenous therapy.

Source: Potter, P. and Perry, A. (1987) *Basic Nursing: Theory and Practice*, p. 827, Mosby, St. Louis.

Why is Rachel L. being treated this way? Is there some connection between alcoholism and her symptoms of low body weight, malnutrition and hypoglycaemia? Part of the answer to questions such as these comes from chemistry.

Clinical note: Rachel's hepatitis involves liver damage. Expected patient outcomes include maintaining body weight within 5% of baseline, blood glucose 100–160 mg/dL, and serum albumin 3.5–5.5 g/dL. Nursing goals include providing adequate nutrition, preventing infection, limiting fatigue and minimising pain.

5.1 METABOLISM

In technical terms, Rachel L. has a metabolic disorder. In her case this disorder is brought about by chronic intake of alcohol, though her admission to hospital this time is due to an acute intake. However, there are other kinds of metabolic disorders that can be inherited, such as phenylketonuria (PKU). In order to understand what is meant by a metabolic disorder, we first need to be clear about what **metabolism** is.

In Chapter 2 we talked in general terms about the idea of homeostasis, or the control of the internal environment. Then chemical homeostasis was discussed at the end of Chapter 4. It was implied there that very many chemical reactions are constantly going on inside the body. The general name for all these chemical reactions, metabolism, is based on the Greek word for 'change'. Since many of these chemical reactions are designed to change a useless chemical into a useful one, to rid the body of a dangerous substance, or to change the size of molecules, you can see that metabolism really is concerned with chemical change. Chemical reactions that are part of the metabolism of the body are referred to as metabolic reactions. Any problem with these metabolic reactions is a metabolic disorder. Metabolic reactions are usually divided into two main types—catabolic and anabolic (see Figure 5.1).

Metabolism
The total of all the chemical reactions that occur in the body. These reactions are also responsible for the heat energy that maintains our body temperature.

FIGURE 5.1 **Metabolic reactions**

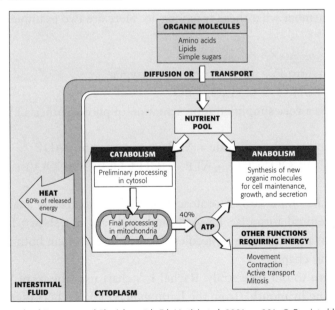

Source: *Fundamentals of Anatomy and Physiology*, 5th Ed, Martini et al, 2001, p. 901. © Reprinted by permission of Pearson Education Inc., Upper Saddle River, N.J.

Catabolic reactions

Reactions that take a large, complex molecule and break it down into smaller, simpler molecules are called **catabolic reactions**. *Cata*, in ancient Greek, meant 'down', which is a useful way of remembering that these reactions break molecules down; or it may be easier to remember that, for the molecule being broken down, it's a *cata*strophe. Here are two examples of catabolic reactions:

alanine (an amino acid) $\rightarrow$ pyruvic acid + ammonia
$$C_2H_4(NH_2)COOH \rightarrow CH_3(CO)COOH + NH_3$$

ethyl alcohol (ethanol) + oxygen $\rightarrow$ carbon dioxide + water
$$C_2H_5OH + 3O_2 \rightarrow 2CO_2 + 3H_2O$$

These examples use chemicals that will be referred to again in this chapter. For now, though, notice how the large reactant molecules have been catabolised into smaller products.

Anabolic reactions

Metabolic reactions that build up larger, more complex molecules from smaller, simpler ones are called **anabolic reactions**. Here, *ana* means 'up', which may help you remember what these reactions do. Here are two examples of anabolic reactions:

carbon dioxide + water $\rightarrow$ glucose + oxygen
$$6CO_2 + 6H_2O \rightarrow C_6H_{12}O_6 + 6O_2$$
(This is a very simplified representation of photosynthesis.)

pyruvate + carbon dioxide + ATP $\rightarrow$ oxaloacetate + ADP
$$CH_3(CO)COO^- + CO_2 + ATP \rightarrow (CH_2COO^-)CO(COO^-) + ADP$$

Pyruvate has nine atoms, oxaloacetate has 11, so this is anabolism. The complete chemical formulae for ATP and ADP are not shown here because they are such large, complex molecules. We will look at them both more closely later on in this chapter.

Let's return to our case study. Rachel L.'s chart uses the term 'NPO'. This means 'nothing by mouth' (from the Latin *nihil per os*). Rachel's digestive system has been stressed by alcohol abuse and cannot perform the mechanical and chemical duties required to catabolise solid food, so she is being given D5W, a 5% dextrose in water solution. Some typical reactions that take place during digestion and the handling of digested material are as follows:

maltose + water → glucose

$$C_{12}H_{22}O_{11} + H_2O \rightarrow 2C_6H_{12}O_6$$

fat + water → fatty acid + glycerol

protein + water → polypeptides

Note that these examples are all hydrolysis reactions (from Chapter 4) and they are all catabolic reactions as well. As you might expect, digestion involves a lot of catabolic reactions, making large molecules small enough to enter the body.

Your studies of the anatomy and physiology of the digestive system might involve some of the chemical aspects of the process. Most of the carbohydrates, fats, proteins and inorganic molecules taken in with the diet have to be broken down. The substances that play a crucial role in bringing about this catabolism are called enzymes, which we'll discuss later in this chapter. Digestion is under the control of various hormones, such as gastrin, which stimulates the release of enzymes when food enters the stomach. Other chemicals are involved too, such as hydrochloric acid (HCl) and various inorganic ions. Examples of the types of chemical reactions involved in digestion are listed as follows:

□ The manufacture and secretion of chemicals essential to catabolism, such as the enzymes mentioned above.
□ The anabolism of nutrients into desired products.
□ The conversion of nutrients from one type (e.g. carbohydrates) into another (e.g. lipids).
□ The storage of nutrients, both organic and inorganic.
□ The inactivation of toxins for later excretion.
□ The homeostatic control of the above (e.g. the careful control of the hydrochloric acid in the stomach).

The chemical reactions referred to here do not all take place in the stomach. All the organs of the digestive system play a part.

5.2 ENZYMES

One aspect of all these chemical reactions that is important to note is the speed with which they happen. After all, some chemical reactions seem to take place very fast, like an explosion, while some are much slower, like the rusting of your car (though some people own cars that seem to rust as fast as an explosion). It's important that the body's chemical reactions take place at the right speed. If you need a sudden supply of adrenaline, for example, the body can't afford to wait around while some difficult chemical reaction plods away, making one adrenaline molecule every 10 minutes. Since most of the biochemical reactions that take place inside the body are very complex and normally happen quite slowly, there must be some way of speeding them up.

That role is performed by molecules called **enzymes**. They act as catalysts, which means they can increase the rate of a chemical reaction by many thousands, sometimes millions, of times, but are not used up in the process. Think of them as like an electric screwdriver—it speeds up the process of assembling a kitchen cupboard, but is not used up itself. Virtually every chemical reaction involved in human metabolism (and there are about 2500 of them) has its own enzyme helping it go faster. Four important things to remember about enzymes are as follows:

Enzymes
Protein molecules that increase the speed of a particular chemical reaction. Their effect is linked to their specific shape.

1. Enzymes are specific; that is, they speed up only one particular chemical reaction.
2. Enzymes are not used up during the reactions they assist.
3. Enzymes can speed up reactions up to several million times.
4. Enzymes are proteins.

Because they are so important, let's look at these points in more detail.

Enzymes are specific

An enzyme usually catalyses only one specific chemical reaction. This helps in naming the enzyme, because the chemist finds what reaction it assists and uses that to label the enzyme. For example, one of the chemical reactions that leads to the formation of urea in the body uses an enzyme called urease. Enzymes always end with -ase. For instance, when maltose changes to glucose, the enzyme responsible is called maltase. The relevant enzyme is often shown written above the arrow of the chemical reaction, like this:

$$\text{maltose + water} \xrightarrow{\text{maltase}} \text{glucose}$$

Enzymes are not used up

The current theory is that enzymes speed up reactions because of their shape. In the example above, maltase has just the right three-dimensional shape to bring the maltose and water close enough together temporarily for the hydrolysis reaction to occur more easily than it would without maltase being present. After the reaction occurs, the resulting glucose is the wrong shape to hang on to the maltase, so it flies off. The maltase is then free to serve as a landing-site for another maltose molecule and another water molecule; see Figure 5.2.

FIGURE 5.2 An enzyme catalysed reaction requires the formation of an enzyme-substrate complex

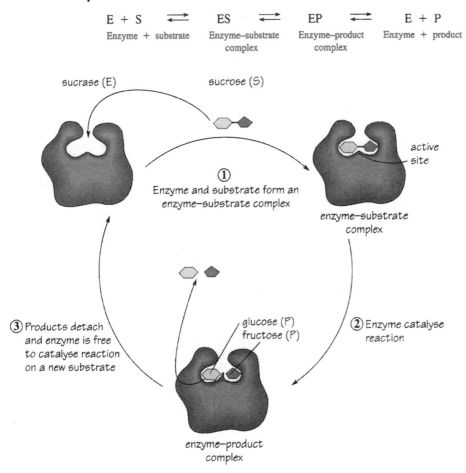

Source: *Principles of Anatomy and Physiology*, 8th Ed, Tortora and Grabowski, p. 47. Reprinted by permission of John Wiley & Sons Inc.

Because maltase, or any enzyme, has just one shape, it can assist only one reaction. This matching of shapes has become known as the lock-and-key model of enzyme action. It assumes that the enzyme's shape is rigid and unchanging. More recent research suggests that some enzymes have the ability to adjust their shape to fit more than one substrate. This has come to be known as the induced-fit model.

Enzymes are very efficient

For instance, one urease molecule can increase the production of urea molecules to 10 000 per second. However, the increase in the rate of other enzyme-assisted reactions can be anywhere from 10^3 to 10^{12} times.

Conditions in the body are important factors in the efficiency of enzymes. After all, the body may not always need an enzyme working at full speed. Since metabolism has to be kept under very fine control, there is a whole range of factors that affect the way enzymes do their job. One important factor is temperature, and another is pH; see Figure 5.3.

It's not too surprising that enzymes work best at normal body temperature—about 36°C. After all, that's the environment in which they are found. Fever or hypothermia can interfere with their function. As well as temperature, the presence of other molecules can significantly affect the action of an enzyme. Some of these molecules are referred to as cofactors. They are often metal ions such as Zn^{2+} or Cu^{2+} and occur near the place on the enzyme where the reaction it is assisting takes place. They help to complete the bonding of smaller molecules, or the breakdown of a larger molecule. The need for these cofactors

FIGURE 5.3 Two factors that affect enzyme function are temperature and pH

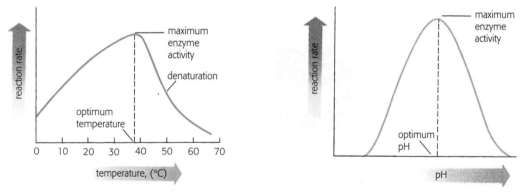

Source: *Chemistry*, 7th Ed, Timberlake, Karen C., 1999, p. 624. © 2001 by Benjamin Cummings. Reprinted by permission of Pearson Education Inc.

is one reason why we need small amounts of these metals in the diet (refer back
to Table 4.2).

Enzymes are proteins

Other molecules, called **coenzymes**, are necessary to make the enzyme in the first
place and they form part of its structure. All enzymes are proteins, but they may
have a non-protein part attached to them. This non-protein part is the coenzyme:

protein + coenzyme = enzyme

Many of the vitamins, such as E, C and all of the B group, are coenzymes, so
their presence in the diet is also crucial for good health.

Another factor that can affect the way enzymes function is the level of acidity
of the environment in which the enzyme is working. Acids and bases will be
discussed in detail later. Keep in mind, for now, that enzymes work best under
the right combination of factors. There are even molecules in the body whose
job is to bind onto the enzyme and prevent it from working, at least temporarily,
as shown in Figure 5.4.

Coenzymes and cofactors
Two of the factors that assist enzymes in their function. Coenzymes are the non-protein part of the enzyme molecule (e.g. vitamin C); cofactors are metal ions (e.g. Zn^{+2}) which occur at the enzyme binding site.

FIGURE 5.4 A competitive inhibitor has a structure very similar to the substrate.
A noncompetitive inhibitor changes the shape of the enzyme, which inhibits its ability to
form a complex.

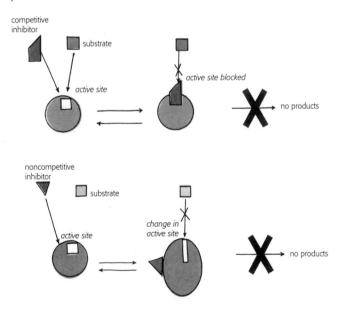

Source: *Chemistry*, 7th Ed, Timberlake, Karen C., 1999, pp. 626, 627. © 2001 by Benjamin Cummings. Reprinted by
permission of Pearson Education Inc.

All these complex processes are part of the delicate task of maintaining homeostasis. The body needs such precise control so it can respond quickly and appropriately to change.

Enzymes and digestion

It's time to put together what we've just said about enzymes, digestion and Rachel L.'s metabolic disorder. Digestion metabolism in general terms can be represented as follows:

$$\text{dietary intake} \rightarrow \text{substrate} \xrightarrow{\text{enzyme}} \text{product}$$
$$\text{(hydrolysis)}$$

The word 'substrate' (see Figure 5.2) refers to the molecules that the enzymes get to work on. The product is the molecule that the enzymes help to produce. If the particular enzyme for the chemical reaction is missing, the result is as follows:

$$\text{dietary intake} \rightarrow \text{substrate} \xrightarrow{\text{no enzyme}} \text{/\!/} \rightarrow \text{little or no product}$$
$$\text{(hydrolysis)}$$

The consequences are twofold. First, the substrate may accumulate in the body. At worst, it may be toxic in large quantities. At best, it may be useless, take a lot of energy to get rid of, and use up the body's resources in doing so.

Second, there is little or no product being produced. This is very serious, especially when that product is crucial to proper body functioning. An example of this is the metabolic disorder referred to as galactosaemia. The normal pattern is:

$$\text{dietary lactose (milk sugar)} \xrightarrow{\text{GPUT}} \text{galactose} \rightarrow \text{glucose}$$

The enzyme that breaks down (catabolises) the galactose to glucose is called galactose-1-phosphate uridyl transferase, which is abbreviated as GPUT. An infant born without GPUT enzyme cannot make glucose from galactose. The galactose accumulates in the body. The excess galactose is, unfortunately, converted by the liver into a chemical called galacticol, which is toxic. As a result, the infant can suffer from cataracts, cirrhosis and mental retardation.

This inherited metabolic disorder affects about one in every 30 000 babies. If it is detected in time, the infant can be placed on a diet low in galactose and the chances of a normal life are quite good. The glucose can be supplied from other sources, so that is not such a problem in this case.

Rachel L.'s case is more complicated. One of the alcohol-related metabolic reactions taking place in her body is shown in Figure 5.5. (Before reading on, is this anabolic or catabolic?) Along with showing the enzymes required for these reactions, the diagram also shows two coenzymes that are necessary—NAD$^+$ and CoA.

NAD$^+$ stands for nicotinamide adenine dinucleotide and CoA simply stands for coenzyme A. These two coenzymes are very important in many reactions in the body. They are mentioned here because if they're too busy dealing with the metabolism of excess alcohol they are not readily available for other reactions. This is one of the clinical problems associated with Rachel L.

FIGURE 5.5 An example of a metabolic alcohol-related reaction in reference to the case study

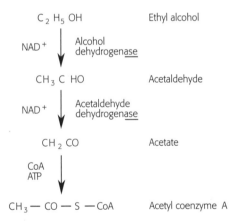

5.3 ENERGY PRODUCTION IN THE BODY

A word we've used a few times without clearly defining it is nutrition. If you chase this word around the dictionary, it seems to be another word for 'sustenance', which doesn't get us very far. In everyday language, nutrition is related in people's minds to eating the right kinds and amounts of food. Figure 5.6 shows this in general terms.

In Chapter 4 the diet of Laurel S. dealt with this to some extent. The 'right' kinds of food are the ones that contain the chemicals essential to do three things:

FIGURE 5.6 Source and destination of nutrients after they have been digested and transported to the liver

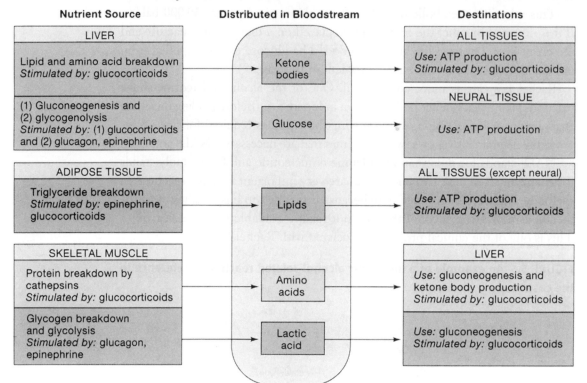

Source: *Fundamentals of Anatomy and Physiology*, 5th Ed, Martini et al, 2001, p. 924. © Reprinted by permission of Pearson Education Inc., Upper Saddle River, N.J.

(1) to supply the materials that make up our cells and tissues; (2) to supply those cells and tissues with the chemicals they need to live, such as water and oxygen; and (3) to provide the energy they require in order to make use of the materials they are given. This section of the chapter looks at the energy supply.

Cells and energy

ATP and ADP
Acronyms for **adenosine triphosphate** (the source of energy used by all cells) and **adenosine diphosphate**.

The energy that the cells use is stored in large molecules called **ATP**, which stands for adenosine triphosphate. How the energy gets into the ATP molecules is the first part of the process and is described in this chapter. How the cell gets that stored energy out and uses it is the second part, described in various places in this book when we look at the function of various cells. At the very beginning, however, is the eating of carbohydrate.

Essentially, all the useful carbohydrate (the 'right' kind) that is taken into the body is digested. That simply means that a series of catabolic hydrolysis reactions take place and reduce almost all of it to glucose, $C_6H_{12}O_6$. Other, more complex carbohydrates play a role as dietary fibre. Depending on the amount of carbohydrate in the diet (the 'right' amount) the glucose is used immediately, stored or (in exceptional circumstances) excreted. The normal glucose level for a healthy adult is 70–100 mg/mL of blood, in between meals. If there is enough excess for it to be stored, it is converted by the liver and skeletal muscles to molecules of glycogen in a process known as **glycogenesis**. Glycogen is a polysaccharide (from Chapter 4), made of a long string of glucose molecules joined together. The reaction can be set out simply as:

$$n(\text{glucose}) \rightarrow (\text{glycogen}) + n(\text{H}_2\text{O}) \ (n \text{ means 'lots of'})$$

When it is needed, the glycogen can be broken down again into glucose (see Figure 5.7).

This reaction, which is the reverse of the one above, is called **glycogenolysis**. (Remember from Chapter 4, *lysis* means to break apart.)

If there is a great deal too much glucose, not all of it can be changed to glycogen, and it is usually converted to fat and stored around the body. If diabetes is present, or if unusual metabolic conditions prevail, some can be excreted in the urine (a condition known as glycosuria).

Whether it comes to the cell straight from digestion, or after glycogenolysis, the glucose is now available to make ATP. Why glucose? Because glucose has a lot of available energy stored in its covalent bonds, put there by plants when they made it with the energy of sunlight. Glucose, in effect, contains stored sunshine. You can recover this energy quite simply yourself by burning a spoonful of sugar; the heat and flame show the release of this stored chemical energy. The body can't use glucose's stored energy directly in this way,

Glycogenesis
The process of converting glucose to glycogen; it occurs in the liver and skeletal muscles.

Glycogenolysis
The process of converting glycogen back into glucose; it also occurs in the liver and skeletal muscles.

FIGURE 5.7 Glycogenesis, glocogenolysis, and the reversible storage of glucose as glycogen

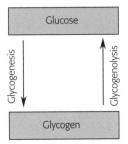

unfortunately. It must be first changed into energy stored in the covalent bonds of ATP.

The name for this changeover from glucose to ATP is **cellular respiration**. It's unfortunate that, in common terminology, 'respiration' also means 'breathing'. The cells are not breathing in that sense, but are using glucose to make the ATP they need for energy. Respiration here has a technical meaning: the use of oxygen to create high-energy ATP molecules out of glucose. Cellular respiration has several steps and we'll take a brief look at each stage. The whole process is quite complex and many readers will not be expected to become familiar with all the details; however it is, at least to some people, a fascinating process.

The production of ATP

In Figure 5.8, a simplified diagram of an ATP molecule, note the three phosphate groups which give it its name: triphosphate. If you removed one of those phosphate groups you would be left with ADP, or adenosine diphosphate. In the body, ATP is made in the following way:

$$ADP + phosphate (P) + energy \rightarrow ATP + H_2O$$

This function can also be shown in a diagram, as in Figure 5.9.

The energy comes mainly from the catabolism of carbohydrates. In the case of Rachel L., with her poor diet and lack of carbohydrate, fats and perhaps proteins are being used to produce the energy; hence her low body weight. This catabolism has to go on all the time, as a cell would use up all its own ATP in less than a minute.

The energy in the glucose covalent bonds is transferred to the ATP in a process called **oxidative phosphorylation**. The word 'oxidative' shows us that oxygen is involved and 'phosphorylation' shows that phosphorous is being added. In brief, the reaction is:

$$coenzyme{-}2H + \tfrac{1}{2}O_2 \rightarrow coenzyme + H_2O + energy$$

FIGURE 5.8 The ATP molecule

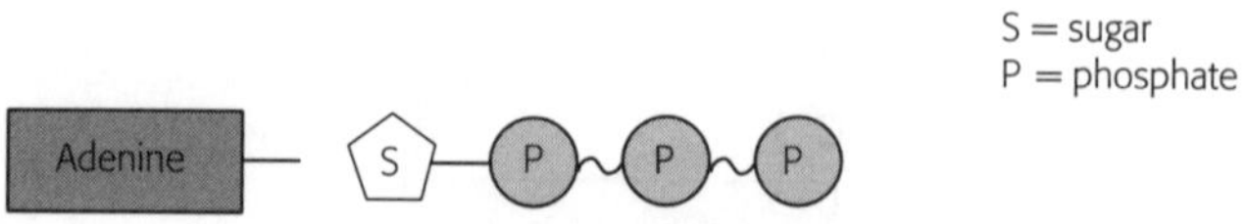

Cellular respiration
The process of converting the energy in carbohydrate or lipid molecules into ATP. There are four steps in this process.

Oxidative phosphorylation
The chemical process by which a phosphate bond is added to an ADP molecule to make a molecule of ATP. The energy for this process is stored in a molecule of glucose.

S = sugar
P = phosphate

FIGURE 5.9 The function of ATP.

Adenosine triphosphate (ATP) is the most important energy currency for cells in the human body. ATP is created by attaching a phosphate group to adenosine diphosphate (ADP); this process requires special enzymes and an energy source. When the high-energy bond joining the third phosphate group to ADP is broken, stored energy is released.

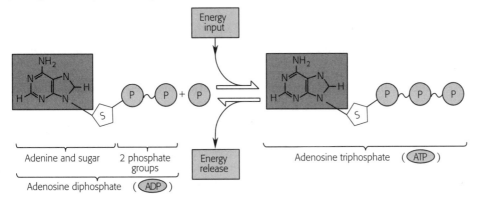

Source: *Fundamentals of Anatomy and Physiology*, 2nd Ed, Martini, F., 1992, p. 102. © Reprinted by permission of Pearson Education Inc., Upper Saddle River, N.J.

The coenzyme gets those vital 2H from a glucose molecule. It then gives those two hydrogens to the oxygen to make water, producing energy at the same time. It's then free to pick up more hydrogen from another glucose molecule and start the process again. The O_2 comes from the oxygen we breathe in with our lungs. Putting the whole thing together, we have:

$$ADP + P + coenzyme–2H + O_2 \rightarrow ATP + coenzyme + H_2O + \tfrac{1}{2}O_2$$

The actual process is a lot more complicated than this general outline, and some interesting medical aspects that relate to cases like Rachel L.'s become clear only if we look at a few of the details. We need, then, to consider one more type of chemical reaction: oxidation–reduction.

Oxidation and reduction reactions

If you've treated a wound, or dyed your hair, with hydrogen peroxide, then you've used a common oxidising agent. What does this mean? An **oxidation reaction** has several chemical definitions, but the ones most useful for us are:

1. There is a loss of electrons.
2. There is a loss of hydrogen or a gain in oxygen.

Oxidation/reduction reactions

Two important types of chemical reactions that occur in the body. Oxidation reactions are those in which (1) there is a loss of electrons and/or (2) there is a loss of hydrogen or a gain in oxygen. Reduction reactions are those in which (1) a gain of electrons and/or (2) a gain of hydrogen or loss of oxygen. These reactions are particularly crucial in cellular respiration.

An example of oxidation, which involves glucose, is shown in the following incomplete chemical reaction:

glucose → glucuronic acid

$$C_6H_{12}O_6 \rightarrow C_6H_{10}O_7$$

There has been a loss of 2H (from 12 to 10) and a gain of 1O (from 6 to 7). That 2H should jog your memory a bit; it gets attached to the coenzyme mentioned previously.

Another, more complex example of oxidation involves a coenzyme that we referred to when discussing Rachel L.'s alcohol metabolism, NAD^+:

$$NADH \rightarrow NAD^+ + 2e^- + H^+$$

There has been a loss of two electrons and a loss of hydrogen. The NADH has been oxidised.

Reduction is the reverse of oxidation:

1. There is a gain of electrons, or
2. a gain of hydrogen, or a loss of oxygen.

If you look at the example of NADH above, it's clear that the NAD^+ has been reduced, because it has gained two electrons and one hydrogen. This example is important here for two reasons. First, this oxidation–reduction reaction is crucial for the whole complex chain of reactions that convert glucose energy to ATP energy; second, as we said before, if the NAD^+ is being used up in alcohol metabolism, there is less available for cell use.

The stages of cellular respiration in more detail

Glycolysis

Glycolysis
One step in cellular respiration, in which glucose is converted to pyruvic acid and 2 ATP molecules are produced.

In this phase, glucose is catabolised through several steps into pyruvic acid (PA); see Figure 5.10.

glucose + 2P (P = phosphate) + 2ADP + $2NAD^+$ → 2PA + 2ATP
+2NADH + $2H^+$ + $2H_2O$

FIGURE 5.10 Glycolysis: a series of biochemical reactions that break down a molecule of glucose into two molecules of pyruvic acid plus energy

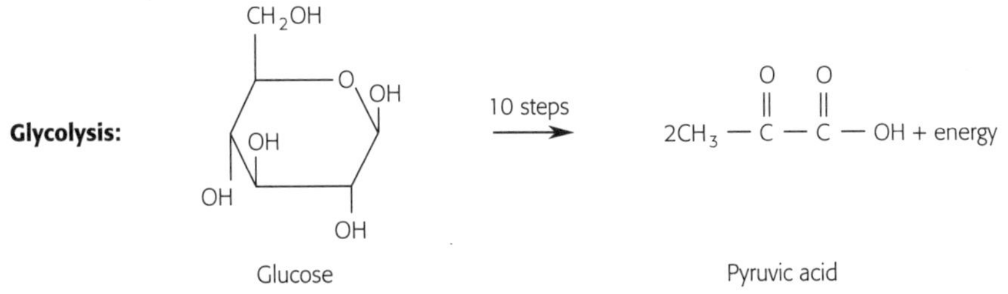

Source: *Essentials of General, Organic and Biological Chemistry*, McMurray, J., 1989, p. 409. © Reprinted by permission of Pearson Education Inc., Upper Saddle River, N.J.

This reaction is catabolic and glucose has been oxidised. There has been a loss of two hydrogens. These are now available for later use.

Pyruvic acid stage

What happens to the pyruvic acid formed above depends on whether or not there is any oxygen around. In anaerobic (no oxygen) conditions, pyruvic acid is reduced to lactic acid, the acid that causes the feeling of fatigue in muscles after exercise. Such anaerobic conditions are common in muscles that are working hard and much anaerobic respiration takes place there. An example of very dangerous anaerobic conditions is a heart attack, where the supply of blood (carrying oxygen) to the heart is cut off. This lack of blood flow is called ischaemia and forces the heart into anaerobic respiration:

$$\text{pyruvic acid} + H^+ + \text{NADH} \rightarrow \text{lactic acid} + \text{NAD}^+$$

The NAD^+ is needed for glycolysis and other metabolic reactions.

If oxygen is present (aerobic conditions), however, the fate of pyruvic acid is quite different. First it loses a CO_2 molecule, leaving behind what is called an acetyl group. The acetyl group attaches to coenzyme A (CoA) and releases two electrons to NAD^+:

$$\text{pyruvic acid} + \text{CoA} + \text{NAD}^+ \rightarrow \text{acetyl-CoA} + CO_2 + \text{NADH}$$
$$CH_3(CO)COOH + \text{CoA} + \text{NAD}^+ \rightarrow CH_3(CO)\text{-CoA} + CO_2 + \text{NADH}$$

In diagram form, the last step can be simplified as shown in Figure 5.11, where the pyruvic acid has been oxidised, losing one hydrogen and two e⁻.

FIGURE 5.11 The oxidation of pyruvic acid under aerobic conditions

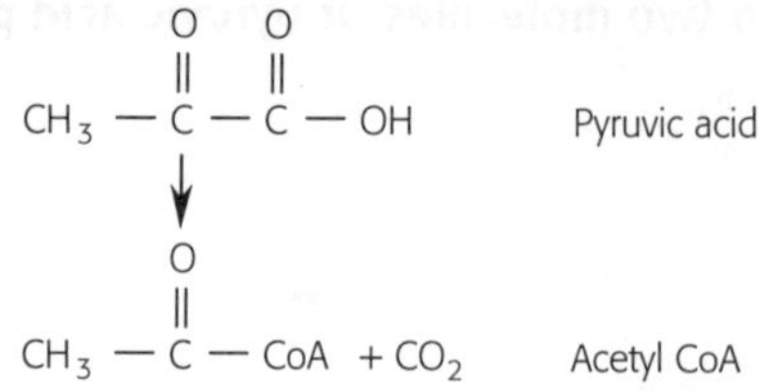

$$CH_3 - \overset{\overset{\displaystyle O}{\|}}{C} - \overset{\overset{\displaystyle O}{\|}}{C} - OH \qquad \text{Pyruvic acid}$$

$$\downarrow$$

$$CH_3 - \overset{\overset{\displaystyle O}{\|}}{C} - CoA + CO_2 \qquad \text{Acetyl CoA}$$

No ATP is made in this phase. The NAD⁺ required for this reaction is regained when the NADH gives up its electrons later on in the part of the process called the **electron-transport chain**.

Electron transport chain
A step in cellular respiration, in which 34 ATP molecules are produced, using hydrogen from the Krebs cycle and oxygen from breathing.

Krebs cycle
A step in cellular respiration. Carbon dioxide and hydrogen are given off and 2 ATP molecules are produced.

Krebs cycle (the citric acid cycle)

This stage takes place within the mitochondrion of the cell, takes nine steps and produces two ATP molecules (see Figure 5.12).

It's called a cycle because the nine steps keep repeating themselves, with a new acetyl group to start it off each time. The acetyl-CoA molecule enters the mitochondrion and combines with oxaloacetic acid. This large, complex molecule then cycles through, until at the end the CoA and oxaloacetic acid are on their own again.

FIGURE 5.12 The Krebs cycle

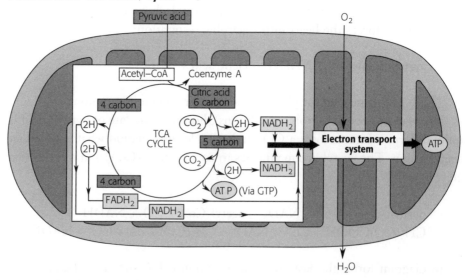

Source: *Fundamentals of Anatomy and Physiology*, 2nd Ed, Martini, F., 1992, p. 104. © Reprinted by permission of Pearson Education Inc., Upper Saddle River, N.J.

The main outcomes of the cycle can be listed as follows:

☐ Two CO_2 molecules are given off. They leave the cell and are exhaled at the lungs.
☐ There are four oxidation reactions among the nine steps. This means that four pairs of 2H atoms (eight atoms in all) leave the cycle (they are used in the next step):

$$8H \rightarrow 8H^+ + 8e^-$$

☐ Three $NADH_2$ molecules are formed from three NAD^+, six H^+ and six e^-. As well, one $FADH_2$ molecule is formed from one FAD, two H^+ and two e^-. Because they have taken in electrons, they are referred to as electron carriers and are very important for the next phase.
☐ Two ATP molecules are made.

Electron-transport phase

This is the main energy-producing part of the whole process. It takes place in the inner mitochondrion and is assisted by structures there called cytochromes. The key element is all those electrons that have been passed on to the $NADH_2$ and $FADH_2$. The main reaction is to transfer those electrons to oxygen atoms, which we breathe in.

$$NADH + 1/2 \ O_2 + H^+ \rightarrow H_2O + NAD^+ + energy$$

As the electrons pass through the cytochromes on their way to combining with oxygen, they make the cytochromes pump H^+ ions across the membrane of the mitochondrion. This leads to a high concentration of positive charge (all those H^+s) on one side of the membrane and a miniature electric battery results. The electric energy from this tiny battery is used to (finally!) join a phosphate group to the ADP to make ATP. In fact, 32 ATP molecules are made during the electron-transport stage. Note too that, if no oxygen was available at this stage, the ATP could not be formed.

Anything that interferes with these steps is potentially life-threatening. For example, cyanide is a poison which acts by combining with the last cytochrome in the sequence, thus preventing it from transferring the electrons to the oxygen. And since, as we said earlier, a hard-won ATP molecule may exist for only a few seconds before being used, the cell would quickly run out of energy and die.

How much energy do we get from all this?

Here's the overall reaction of cellular respiration:

$$\text{Glucose} + 36\text{ADP} + 36\text{P} + 36\text{H}^+ + 6\text{O}_2 \rightarrow 6\text{CO}_2 + 36\text{ATP} + 42\text{H}_2\text{O}$$

These 36 ATP molecules represent about 1.1×10^6 J of energy. If we had simply burned the sugar in a spoon, like this:

$$\text{glucose} + 6\text{O}_2 \rightarrow 6\text{CO}_2 + 6\text{H}_2\text{O}$$

the total energy available would have been about 2.9×10^6 J. Therefore, cellular respiration is about 38% efficient, which, as we said in Chapter 4, is better than most cars. The important thing is, though, that the oxidation of glucose is done by the body in a very controlled way.

Burning sugar in a spoon releases its energy so quickly that a great deal of heat is generated very rapidly. Such heat would, of course, damage the body. Thanks to cellular respiration and the careful control of it by enzymes, we can burn glucose at body temperature. In fact, the remaining 62%, which did not go into making ATP, appears as heat energy which helps keeps the body warm, or which must be disposed of, as we saw in Chapter 2.

Metabolism of lipids for energy

We said earlier that Rachel L. was out of glucose and appeared to be using her body fat as a source of fuel. ATP is no good as a store of energy, because it gets used up too quickly. The body uses glycogen as a short-term store of glucose, but when this is gone, fats can then be used. This is a normal part of body metabolism and happens all the time in the body anyway, especially in the heart and liver. At rest, the body gets about half of its energy from glucose and half from fats. When we exercise, glucose is the readily available, preferred fuel. When a fat—for example, a triglyceride—is broken down by hydrolysis, the products are one molecule of glycerol and three molecules of fatty acid (Chapter 4 discussed fats).

$$\text{triglyceride fat} + 3\text{H}_2\text{O} \rightarrow \text{glycerol} + 3 \text{ fatty acids}$$

It can be represented as shown in Figure 5.13.

The glycerol part can be converted to pyruvic acid, which then ends up in

FIGURE 5.13 An overview of lipid catabolism

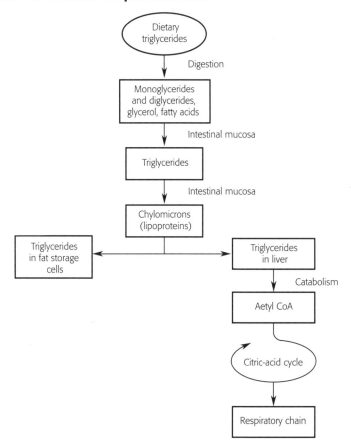

the Krebs, or citric acid, cycle. The fatty acid part gets broken down into pieces
which are two carbon atoms long. This is called beta oxidation and takes place
in the mitochondria.

Imagine starting with a triglyceride that had 18 carbon atoms in it. After one
beta oxidation, there would be a piece with 16 carbon atoms and one with two
carbon atoms. The two-carbon molecule forms acetyl-CoA. The benefits of this
process are as follows:

☐ 12 ATP from the acetyl-CoA in the Krebs cycle
☐ five ATP from NADH and $FADH_2$ produced during the beta oxidation
☐ any gains from the glycerol → pyruvic acid change.

From one 18-carbon fatty acid molecule, a total of 148 ATP are available. This is almost 1.5 times as much as from three glucose molecules. The reason we don't use these fatty acids as much as glucose seems to be that they are insoluble in water and thus harder for the body to use.

Metabolism of proteins for energy

Protein is needed for so many other things that it's the fuel of last resort, but during malnutrition the body may have to rely on stores of protein for energy. The process begins with **deamination**, the removal of amine groups from the amino acids. This takes place in the liver and kidneys. For example, the deamination of alanine is shown as follows:

Deamination

The removal of amine groups (those containing nitrogen) from amino acids. It is a step in the breakdown of protein for energy purposes.

alanine → pyruvic acid (a keto acid) + ammonia

The ammonia is excreted as urea through a reaction mentioned in Chapter 4:

ammonia + carbon dioxide → urea + water
$$2NH_3 + CO_2 \rightarrow NH_2CONH_2 + H_2O$$

This last step can be illustrated by Figure 5.14.

Pyruvic acid, of course, is now quite familiar to us. Not all amino acids produce pyruvic acid when broken down, but six of the 20 do; three others produce acetyl-CoA, two produce oxaloacetic acid for the Krebs cycle, five produce acetoacetyl-CoA which can produce acetyl-CoA and so forth.

FIGURE 5.14 The urea cycle takes two metabolic waste products—carbon dioxide and ammonia—and produces a molecule of urea.

Urea is a relatively harmless, soluble compound that is excreted in the urine.

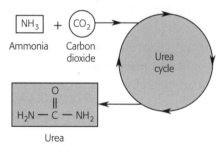

Source: *Fundamentals of Anatomy and Physiology*, 2nd Ed, Martini, F., 1992, p. 835. © Reprinted by permission of Pearson Education Inc., Upper Saddle River, N.J.

Energy production and Rachel L.

Alcoholism brings about many disruptions to normal metabolism. In the case of energy production, we now have enough background information to understand more clearly Rachel L.'s condition. The problem is summarised as follows:

1. Ethyl alcohol (ethanol) (C_2H_5OH) is an energy source and can be used to make ATP. Some of the chemical reactions that make this possible were shown earlier in this chapter. The alcohol is converted in only three steps to acetyl-CoA, which is used in the Krebs cycle. When the body is healthy, it can metabolise anywhere from 7 g to 18 g of alcohol per hour. This use of alcohol as a fuel can be a danger. Accumulation of fat within the liver, due to the use of alcohol for energy rather than the fatty acids normally used, is a real problem.

2. The metabolism of alcohol uses up essential NAD^+ and this has at least two damaging consequences. First, there is an increase in the amount of lactic acid in the blood. Remember, lactic acid is a product of anaerobic cellular respiration and is responsible for the feeling of muscle fatigue. There is evidence that lactic acid also brings on a clinical condition called gout. Second, the loss of available NAD^+ can lead to hypoglycaemia (one of Rachel L.'s conditions), which means lowered levels of glucose in the blood, especially true for alcoholics with poor diets.

   Since NAD^+ is made from Vitamin B_3, or nicotinic acid, the alcoholic has a greater need for it. If enough quantities are not present in the diet, vitamin deficiency may occur. When not severe, vitamin deficiency symptoms can include dermatitis, glossitis and diarrhoea.

3. Similarly, some of the cofactors, such as NAD^+, of other enzymes required for alcohol metabolism may be affected. Metal ions such as Mg^{2+}, K^+ and Zn^{2+} may be lost through the urine.

Summary

The information in this chapter is perhaps best summarised by referring to our two case studies, Laurel S. (Chapter 4) and Rachel L. Laurel S. is a diabetic, so her problem is a lack of insulin. We need, then, to see what role insulin plays in energy production. There are five main factors involved:

1. Insulin helps the transport of glucose into the cells of the body. Remember, that's where cellular respiration takes place.

2. Once inside the cell, insulin speeds up the oxidation of glucose. Recall an earlier oxidation reaction:

glucose → glucuronic acid

This is one important step in the catabolising of glucose.

3. Insulin assists in glycogenesis. One of the steps is shown:

$$\underset{\text{insulin}}{\overset{\text{glucinase}}{\text{glucose} + \text{ATP} \rightarrow \text{glucose–6–P} + \text{ADP}}}$$

4. Insulin slows down glycogenolysis in the liver. Glycogenolysis is the breakdown of glycogen to glucose.

5. Insulin promotes lipogenesis, the manufacture of lipids from glucose.

If you look carefully at each of these five effects of insulin, you will see that they all have the same result—to remove glucose from the blood supply. At the same time, insulin assists cellular respiration, either helping to make the glucose available at the right place, or acting as an enzyme in some of the key steps. A deficiency in insulin leads to increased glucose in the blood (known as hyperglycaemia) and in the urine (glycosuria). Of course, there are other serious consequences of diabetes, but many of them are directly related to lack of control over glucose and therefore lack of control over energy production.

Hopefully we now understand something of Rachel L.'s metabolic disorders. The excessive intake of alcohol has badly disrupted her chemical homeostasis. The diet of an alcoholic is often poor and often leads to deficiencies in essential nutrients. The operation of the digestive system is impaired, so whatever is eaten is not metabolised efficiently. The cofactors and coenzymes essential for metabolism control are lacking. Supplies of glucose for energy are in short supply and in serious cases must be given intravenously. The process of cellular respiration is impaired because alcohol has replaced glucose as the fuel and has removed important chemicals from the several stages of the process. All this—not to mention the toxic effects of alcohol on the brain and nervous system, possible cirrhosis of the liver and damage to the lining of the stomach.

The treatment of both Laurel S. and Rachel L. includes careful monitoring of their diets. The amount and kind of chemicals entering their bodies profoundly influences the complex, essential and interrelated chemical reactions of metabolism.

Questions

Level 1

1. What is glycogenesis? Where does it occur?
2. Which monosaccharide is the principal one remaining in the bloodstream after passing through the liver?
3. What happens to excess carbohydrate that cannot be immediately used or converted to glycogen?
4. What is glycosuria?
5. What types of compounds does the body use to store energy?
6. Name two pathways in which ADP may be converted to ATP.
7. How is glucose removed from the bloodstream? What role does insulin play in this process?
8. Outline the process by which ADP is converted to ATP.
9. What is glycolysis? Is it an aerobic or anaerobic process?
10. Define the term 'enzyme'. What are some of the factors that assist its function?
11. How much energy is produced in the cellular respiration of one molecule of glucose?
12. What are the products of the catabolism of a lipid when it is used for energy production?

Level 2

13. What are the major chemical imbalances associated with Rachel L.'s alcoholism?
14. Explain briefly what happens in the Krebs cycle.
15. Where does the electron transport chain operate? Briefly describe what happens there.
16. Give an example of an oxidation reaction. Show which molecules are oxidised. Write an equation to show the oxidation of NADH.
17. Discuss in general terms the possible effects of a lack of the enzyme lipase on the digestion of lipids.

Level 3

18. Show the interrelationship of glycolysis, the Krebs cycle, glycogenesis and glyco-genolysis.
19. Why are enzymes crucial to life? Name and discuss several factors that control enzyme activity.
20. Explain what is meant by enzyme specificity. Illustrate your answer with the lock-and-key model.

21. Suppose a body-builder wished to add muscle bulk and decided to try a high-protein diet. Explain why this would or would not work.
22. Name a potential consequence of copper deficiency.

CALCULATION PROBLEM RELATED TO THE CASE STUDY

Rachel L. is receiving dextrose IV. Suppose she is to receive half a litre of dextrose 5% over 4 hours. The giving set delivers 20 drops/mL. What is the required drip rate in drops/min?

Answer:

rate (drops/min) = volume (drops)/time (mins)

rate = volume/time (hours) × 60

rate = 500 mL × 20 drops/mL/4 hours × 60

rate = 500 × 20/4 × 60 drops/min

rate = 125/3 drops/min = 41.66 = 42 drops/min.

6

THE RESPIRATORY WARD AND CLIENT CARE

THE GAS LAWS, SOLUBILITY AND DIFFUSION OF GASES

Chapter outline

6.1 **Ventilation.** Your objectives: to understand how gases get into and out of the lungs; and to know and apply the concept of pressure along with the principle gas laws.

6.2 **Solubility of gases in liquids.** Your objective: to understand the concept of solubility, specifically as it applies to gases dissolved in the blood.

6.3 **Diffusion.** Your objective: to understand and apply the principles of the movements of substances across cellular membranes.

6.4 **Air pollution.** Your objective: to apply the information and concepts of this chapter to understanding some of the health-related problems of air pollutants.

INTRODUCTION

As we have seen in earlier chapters, maintaining a constant internal environment is not easy. For example, the cells deep within the body are completely isolated from the atmospheric oxygen which they need, and therefore must be supplied by other means. The fascinating way in which our bodies solve that particular problem is the subject of this chapter. The nurse normally sees this aspect of body function only when something goes wrong with it. But frequent monitoring of a patient's breathing rate, or 'resps', is a common nursing task. Why are we so interested in how many breaths our patients take in a minute? We'll also be talking about some of the more common health problems associated with the respiratory system.

Case Study: Mr Vern W.

Chronic obstructive pulmonary disease (COPD) refers to those conditions that hinder the flow of air into and out of the lungs. At 71 years of age, Mr W. suffers from one of these conditions, known as emphysema. As this part of your clinical training concerns looking at oxygen therapy, your clinical supervisor has supplied you with the following extra information about Mr W.'s condition and treatment:

Mr W. is a long-term victim of panlobar emphysema. In general terms, he has three main problems. First, he suffers from increased *lung capacity*, due to the joining together of his alveoli. Second, there is increased *airway resistance*, so more effort is needed to breathe. Third, there is altered O_2/CO_2 *exchange* in the lungs. Part of his treatment consists of lifestyle changes and medication. His oxygen therapy calls for him to be placed on a Venturi mask, 25% oxygen. This mask delivers low oxygen concentrations to avoid the possible loss of hypoxic respiratory drive.

Clinical notes: Nurses are often asked to consult with the physician about the criteria used to determine when ventilatory support is needed. For example, if P_{aO2} is less than 60 mm Hg, or when P_{aCO2} is greater than 50–60 mm Hg, which is more than 10 mm Hg above normal.

These nursing and clinical notes refer, in technical language, to respiration in humans. As we said in Chapter 5, cellular respiration refers to the cells' use of oxygen to generate energy. Here we are looking at the more common meaning of the word 'respiration'—how oxygen and other gases get into and out of the body. This process is commonly referred to as breathing. There are four important points to keep in mind about breathing, three of which we'll talk about in this chapter. They are ventilation, moving air into and out of the lungs; diffusion, moving the required gases in the air from lungs to bloodstream; transport of those gases to the cells; and the control of the rate of respiration (not discussed in this text). Let's start by looking at ventilation and thinking of the lungs as balloons.

6.1 VENTILATION

'**Ventilate**' is a word from Latin meaning 'wind blowing'. Its connection with breathing is pretty clear. To get any wind to blow, there must be a difference in air pressure. Pressure, remember, is force/area. Where there is a lot of air, as in a high pressure system, there is a greater force due to the weight of that air pressing down on the surface of the earth. If there is an area of lower pressure nearby, the air will flow into it from the high pressure area. That flow is, of course, what we call wind.

Similarly, if we want air to flow into the lungs, we have to make sure that the pressure inside the lungs is lower than the pressure outside. Our lungs become a miniature low pressure system, therefore, when we inhale, or inspire. To get air out of the lungs, we have to make the pressure inside greater than outside so that the air flows out like a brief, tiny windy day. That is, we exhale, or expire. Figure 6.1 demonstrates the process.

Any good anatomy text will show you a reasonable photograph or diagram of the lungs. Figure 6.2 shows you the pressure changes as air moves in and out of them.

1. *In-between breaths*. The pressure inside the lungs is the same as the pressure of the atmosphere—approximately 101.3 kilopascals (kPa), or 760 mm Hg.
2. *Inhalation (inspiration)*. The size (volume) of the lung increases. This is caused mainly by the contraction of the diaphragm, which forces the contents of the abdomen downwards, leaving a little space into which the lungs can expand.

Ventilation
Commonly called breathing; the movement of air into and out of the lungs by inhaling (inspiring) and exhaling (expiring).

FIGURE 6.1 Pressure and volume relationships in the lungs.

(a) At the start of inspiration the pressure in the lungs is identical to that of the surrounding atmosphere. No net movement of air occurs under these conditions. (b) As inspiration commences, the volume of the thoracic cavity increases, so the pressure of the lungs decreases. Air then moves into the lungs. (c) During expiration the volume of the thoracic cavity decreases. Pressure inside the thoracic cavity then exceeds that of the surrounding atmosphere, forcing air out of the lungs.

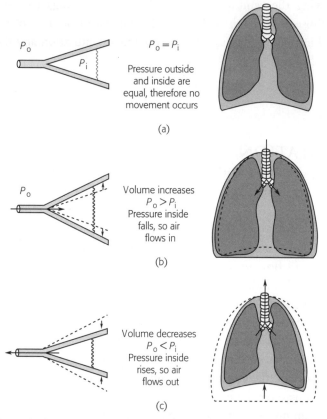

Source: *Fundamentals of Anatomy and Physiology*, 2nd Ed, Martini, F., 1992, p. 764. © Reprinted by permission of Pearson Education Inc., Upper Saddle River, N.J.

When this happens, there is the same amount of air in a larger space, which means lower pressure. The pressure inside the lung is now only 100.3 kPa or 752.5 mm Hg. Therefore, air flows in from the outside. It is common to refer to the pressure inside the lung at this point as negative pressure, because it is 1 kPa (or 7.5 mm Hg) less than the atmospheric pressure. The amount of air that enters the lung under normal inhalation is referred to as the tidal volume

FIGURE 6.2 Pressure changes during inspiration and expiration.

These graphs follow changes in the intrapleural and intrapulmonary pressures during a single respiratory cycle, and relate these changes to the tidal volume.

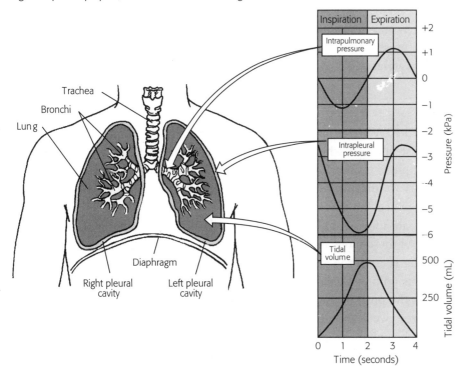

Source: *Fundamentals of Anatomy and Physiology*, 2nd Ed, Martini, F., 1992, p. 765. © Reprinted by permission of Pearson Education Inc., Upper Saddle River, N.J.

(TV); see Figure 6.2 and Table 6.1. For a healthy young adult, it's about 500 mL (which you may see in older units, i.e. 500 cc) of air.

3. *Exhalation (expiration)*. The volume of the lung decreases. Normally this is due only to the lungs collapsing slightly inwards, like a deflating balloon, because they are elastic. We don't need to use any muscles actively to expire, unless we want to breathe more forcefully. The extra tidal volume of 500 mL of air is now squeezed into a smaller space. The pressure rises to about 102.3 kPa or 767.5 mm Hg. Since this is higher than atmospheric pressure, the air flows out into the atmosphere. This extra pressure is referred to as positive pressure, in this case 1 kPa higher than atmospheric pressure.

If the lungs are elastic and tend to deflate like a balloon, what stops them from totally collapsing? Obviously, they must be pulled from the outside.

In between the lungs and the thoracic cavity is a space called the pleural cavity. It is full of fluid at a negative pressure of between –8 kPa (during inhalation) and –4 kPa (during exhalation). This negative pressure acts as a suction force to hold the lungs against the chest wall. If the pleural space is opened to the outside, for example by an accident such as a knife puncture, air rushes in. There is no more negative pressure holding the lungs open and they can collapse, a condition known as atelectasis.

The gas laws

We've just said that when the lungs expand the pressure inside them drops. In our example of inspiration above, the decrease was from 101.3 kPa to 100.3 kPa. Why? Try to imagine what is happening inside a balloon filled with air. It's not easy to picture, but there are a great many air molecules inside an average-sized balloon, all moving at tremendous speeds—up to hundreds of metres per second. As these molecules hit the inside of the balloon's skin, they push it out, expanding it, until the elastic pull of the rubber equals the outward push of the air. At this point we have an example of equilibrium.

Since the push of the air is the force and the size of the skin of the balloon is the area, we have an example of air pressure, or force/area. Now, if you increase the volume of the balloon, without adding more air, making the area of skin larger, the pressure inside the balloon gets smaller; that is:

$$\text{if } P_1 = 4 \text{ N/2 m}^2 = 2 \text{ Pa, then } P_2 = 4 \text{ N/4 m}^2 = 1 \text{ Pa}$$
$$\text{(small area)} \qquad\qquad \text{(larger area)}$$

Of course, increasing the size of the balloon by expanding it changes the volume as well as the area. This change in pressure with a change in area or volume has been written by physicists as a law, called **Boyle's Law**; see Figure 6.3.

Boyle's Law
The pressure of a gas is inversely proportional to the volume it is confined in, providing the temperature remains unchanged.

Boyle's Law says that the pressure is inversely proportional to volume, as long as the temperature doesn't change too. As the volume gets larger, the pressure gets smaller; and as the volume gets smaller, the pressure gets larger. Mathematically, Boyle's Law is written $V \propto 1/P$ ($\propto$ means 'is proportional to') for constant temperature.

If you prefer mathematical equations with = signs instead, we have $P \times V = k$ (k stands for an unchanging number called a constant). The value of k depends

FIGURE 6.3 Boyle's Law: decreasing the volume of a gas sample increases crowding of the molecules and thereby increases the pressure.

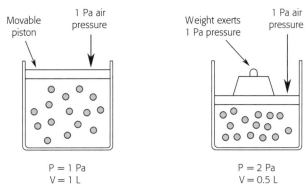

Source: *Essentials of General, Organic and Biological Chemistry*, McMurry, J., 1989, p. 99. © Reprinted by permission of Pearson Education Inc., Upper Saddle River, N.J.

on the particular situation being looked at (e.g. what sort of balloon it is). Here's an example of this equation at work:

$$100 \text{ Pa} \times 5 \text{ cm}^3 = 500 \text{ Pa.cm}^3$$

Here, the 500 Pa.cm³ is the constant, which must remain unchanged. Now, double the volume.

$$? \text{ Pa} \times 10 \text{ cm}^3 = 500 \text{ Pa.cm}^3$$
$$? \text{ Pa} = 500/10 = 50 \text{ Pa}$$

Doubling the volume means the pressure goes down by half.

But what if the temperature does change? Well, temperature is related to heat energy. Since the heat energy of a molecule is really its energy of motion, when you heat up a gas the molecules move faster. The faster they move, the harder they slam into the walls of the balloon, so that there is an increase in force and therefore an increase in pressure. That is:

$$\text{if } P_1 = 4 \text{ N/2 m}^2 = 2 \text{ Pa, then } P_2 = 8 \text{ N/2 m}^2 = 4 \text{ Pa}$$
$$\text{(small force)} \qquad\qquad \text{(larger force)}$$

This change in pressure with a change in temperature is also expressed as a law, called **Charles' Law**. It says that volume is directly proportional to temperature as long as the pressure doesn't change. Mathematically: $V \propto T$, or

Charles' Law
The volume of a gas is directly proportional to the temperature, in Kelvin, providing the pressure remains unchanged.

$V/T = k$ with constant pressure. As the temperature goes up, so does the volume. And if the temperature drops, say by half, so does the volume, also by half. Strictly speaking, the temperature should be expressed with the SI unit of temperature, the Kelvin. (Just add 273 to the Celsius temperature to change it to Kelvin.)

Sometimes it's easier to understand these changes in volume and temperature if we think about the situation before the change and after the change.

$$\text{Before: } T_1/V_1 = k \qquad \text{After: } T_2/V_2 = k$$

This means that:

$$T_1/V_1 = T_2/V_2$$

We can put Boyle's Law together with Charles' Law to cover situations where pressure, volume and temperature may all be changing. The result is called the **General Gas Law**:

$$P_1V_1/T_1 = P_2V_2/T_2$$

For Mr W., for example, let

$P_1 = 100$ kPa	(starting to inhale)	$P_2 = x$ kPa	(exhaling)
$T_1 = 33°C = 306$ K	(mask oxygen)	$T_2 = 36°C = 309$ K	(warmed by body)
$V_1 = 500$ mL	(tidal volume)	$V_2 = 500$ mL	$(\text{Vol}_{in} = \text{Vol}_{out})$

We now rearrange the general gas law equation to find P_2:

$$P_2 = P_1V_1T_2/T_1V_2 = 100 \text{ Pa} \times 500 \text{ mL} \times 309 \text{ K}/306 \text{ K} \times 500 \text{ mL}$$
$$P_2 = 101 \text{ kPa}$$

The slight warming of the air by his lungs led to a slight increase in the pressure of air exhaled.

Lung volumes and capacities

Although breathing is normally automatic, we have some conscious control over both the rate we breathe at and the amount of air that enters and leaves the lungs.

General Gas Law
Combines Boyle's and Charles' Laws into one expression. Mathematically, for a given sample of a gas $P_1V_1/T_1 = P_2V_2/T_2$. Here, P1 means the pressure of the gas before any changes were made and P2 is the pressure after changes occurred; likewise for V and T

That is, we are not limited to that 500 mL of tidal volume per breath, nor to the normal 12 breaths per minute. The lungs in fact hold close to 6000 mL or 6 litres of air. If we breathe in only 500 mL, we must account for the rest. Table 6.1 shows how physiologists classify the amounts of air involved in pulmonary ventilation.

This information can also be presented graphically as respiratory performance and volume relationships; see Figure 6.4.

TABLE 6.1 PULMONARY VOLUMES AND CAPACITIES (figures for adult male)

Pulmonary Volumes	Pulmonary Capacities
Tidal volume (TV): amount normally inspired and expired; approx. 500 mL	Inspiratory capacity (IC): TV + IRV (4100 mL)
Inspiratory reserve volume (IRV): extra that can be inspired with effort; approx. 3600 mL	Functional residual capacity (FRC): ERV + RV (2400 mL)
Expiratory reserve volume (ERV): extra that can be expired after normal expiration; approx. 1200 mL	Vital capacity (VC): TV + IRV + ERV (5300 mL)
Residual volume (RV): approx. 1200 mL	Total lung capacity (TLC): approx. 6500 mL

FIGURE 6.4 Respiratory performance and volume relationships

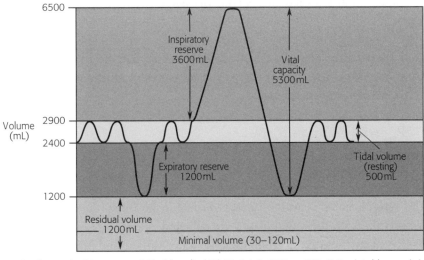

Source: *Fundamentals of Anatomy and Physiology*, 2nd Ed, Martini, F., 1992, p. 768. © Reprinted by permission of Pearson Education Inc., Upper Saddle River, N.J.

From Table 6.1 or Figure 6.4 you can work out that the vital capacity, for example, is the amount of air that you can take in and breathe out with maximum effort. The vital capacity of professional athletes and operatic singers is greater than the normal value shown above, because of training and exercise.

Let's go back to our case study. Mr W.'s vital capacity, by comparison, may be as low as 2300 mL; half the normal value. In severe cases it may drop even lower. Mr W.'s vital capacity is low because his lungs have lost a great deal of their elasticity. They no longer contract, so their volume does not decrease and the pressure does not rise. To breathe out, he has to use his intercostal and abdominal muscles to squeeze the lungs. This is very tiring and can become painful.

6.2 SOLUBILITY OF GASES IN LIQUIDS

Another interesting feature of Table 6.1 is the residual volume. Even with maximum expiratory effort, approximately 1200 mL of air remain in the lungs. (This is not the same air all the time, however.) As you might guess, this is not mere poor planning on the part of the respiratory system. However, to see the importance of this 'dead' air we need to examine what happens to the gases in the air when they finally get into the alveoli.

It's time to stop talking about 'gases' and 'air' and be more specific. The atmosphere is composed of several gases, the most important for us being oxygen (O_2), carbon dioxide (CO_2), nitrogen (N_2) and water vapour (H_2O). Each of these contributes its pressure to the total atmospheric pressure of 101.3 kPa. This contribution is called the **partial pressure** (written p_{N_2}, for example) of that gas.

Partial pressure
The pressure exerted by a specific gas, as a contribution to the total pressure of all the gases present. For a gas X, it is written p_x pascals.

Once again, this fact can be expressed in the form of a law, called Dalton's Law of Partial Pressures. In effect, this law states that the total pressure of a mixture of gases is the sum of the partial pressures of all the gases present. In symbols, we can express this as:

$$P_{total} = p_{gas\ 1} + p_{gas\ 2} + p_{gas\ 3} + \ldots$$

There are two things to say about this law here. First, the pressure inside the body is equal to the outside atmospheric pressure. Second, if a new gas is added to the system, such as water vapour during inspiration, then all the partial pressure values of the other gases are affected. Table 6.2 shows the partial pressure values of oxygen, carbon dioxide, water vapour and nitrogen at various places in the body.

TABLE 6.2 PARTIAL PRESSURES OF GASES AS THEY ENTER AND LEAVE THE LUNGS

Place	Gas	Partial pressure	(mm Hg & kPa)	% of total
Atmosphere	O_2	159 mm Hg	21.2 kPa	20.93
	CO_2	0.3 mm Hg	0.04 kPa	0.04
	H_2O	3.7 mm Hg	0.49 kPa	0.48
	N_2	597 mm Hg	79.6 kPa	78.58
Total		760 mm Hg	101.3 kPa	100.0
Alveoli	O_2	104 mm Hg	13.9 kPa	13.7
	CO_2	40 mm Hg	5.33 kPa	5.26
	H_2O	47 mm Hg	6.26 kPa	6.18
	N_2	569 mm Hg	75.8 kPa	74.82
Total		760 mm Hg	101.3 kPa	100.0
Expired	O_2	120 mm Hg	16.0 kPa	15.79
	CO_2	27 mm Hg	3.59 kPa	3.54
	H_2O	47 mm Hg	6.26 kPa	6.18
	N_2	566 mm Hg	75.4 kPa	74.43
Total		760 mm Hg	101.3 kPa	100.0

As you can see, the composition of the gases inspired and expired changes quite a bit. The amount of oxygen is depleted as it is used by the cells and carbon dioxide and water vapour are increased. The changeover takes place in the alveoli and is discussed shortly.

The reason for the residual volume (RV) mentioned earlier can now be made clearer. If all the air inspired was taken straight to the alveoli and then flushed out again, the changeover of gases would happen too quickly. The residual volume acts to slow down any sudden changes in the partial pressures. This may seem to answer one question only to raise another: why is a rapid changeover of gases a problem? That question takes us deep into an individual alveolus to see what happens to the gases when they finally arrive there.

Gases can dissolve in water, as any admirer of beer or champagne knows well. To dissolve a solid such as salt or sugar, we usually stir it in. With gases, we usually bubble them through. How much of the gas dissolves out of the bubble into the water (or champagne) depends on three things. The first is the partial pressure of the gas. This makes sense, because the higher the partial pressure, the more of the gas there is. But some gases are insoluble in water. Just as you can add sand to water all day long without dissolving a single grain, the same is true of some gases. Some dissolve very well, some very poorly. The second factor, then, is called the **solubility coefficient** of the gas. It's a measure of how easily a gas dissolves in water of a known temperature. The larger the coefficient, the more soluble the gas is; see Table 6.3.

The third factor is the temperature of the water. Many gases dissolve best in cold water. But we won't worry about temperature here, because we can safely assume that the temperature of water in the body does not change from around 36°C.

This means, for example, that 2.4×10^{-4} mL (0.000 24 mL) of oxygen will dissolve in every available 1 mL of H_2O for each 1 kPa of oxygen partial pressure, providing the temperature doesn't change. In Table 6.3, the temperature of the water is 0°C.

The partial pressure and the solubility coefficient combine in another (the last!) of our gas laws, **Henry's Law**. The amount (volume) of gas that will dissolve in a liquid such as water is proportional to its partial pressure (p_{gas}) and its solubility coefficient (c):

$$V = p_{gas} \times c$$

Figure 6.5 presents this in diagram form.

Let's do an example. From Table 6.2 we have the partial pressure of oxygen in the alveoli as 13.8 kPa.

volume = partial pressure × coefficient of solubility
volume = 13.8 kPa × 0.000 24 mL/mL H_2O/kPa
= 0.0033
= 3.3×10^{-3} mL/mL H_2O

TABLE 6.3 SELECTED SOLUBILITY COEFFICIENTS

Gas	mm Hg ($\times 10^{-5}$)	kPa ($\times 10^{-4}$)
Oxygen	3.2	2.4
Carbon dioxide	75	56
Nitrogen	1.6	1.2

FIGURE 6.5 Henry's Law and the relationship between solubility and pressure.
(a) A solution containing dissolved gas molecules at equilibrium with air under a given pressure. (b) Increasing the pressure drives additional gas into solution until a new equilibrium becomes established. (c) When the pressure decreases, some of the dissolved gas molecules leave the solution until equilibrium is restored.

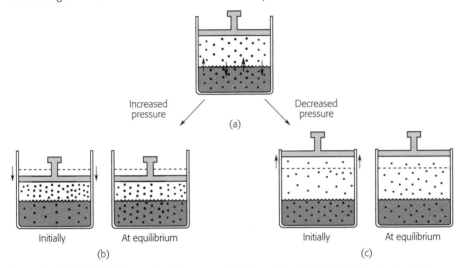

Increased pressure   Decreased pressure
(a)

Initially   At equilibrium   Initially   At equilibrium
(b)   (c)

Source: Martini, F. (1992) *Fundamentals of Anatomy and Physiology*, 2nd edn, Prentice Hall, New Jersey, p. 770.

Summary of the gas laws

Boyle's Law: if the temperature remains unchanged, then

$$P \times V = \text{constant}$$

Boyle's Law helps explain the movements of gases into and out of the lungs. As the lung volume increases during inspiration, the pressure decreases below atmospheric pressure and air enters the lung; when the lung volume decreases during expiration, the pressure rises above atmospheric pressure and air leaves the lung.

Charles' Law: if the pressure remains unchanged, then

$$V/T = \text{constant}$$

Normally, temperature changes are relatively small during breathing, but air slightly expands as it is warmed by the body.

Dalton's Law of Partial Pressures:

$$P_{total} = P_{gas\ 1} + P_{gas\ 2} + P_{gas\ 3} + \ldots$$

This law helps us determine the direction of gas flow in the body and also tells us where we can expect gas concentrations to be highest or lowest.

Henry's Law: the amount of gas that can dissolve in a solvent at a specified temperature is proportional to the partial pressure of that gas and its solubility coefficient, assuming the temperature is constant.

Henry's Law tells us how much gas we can expect to find dissolved in the body's fluids at any given place.

6.3 DIFFUSION

From what has been said about gases and solubility it might sound as though the alveoli were full of water for the gases to dissolve in, but this is not the case. Only the inner surface of each alveolus—when the lung is healthy—is lined with a thin layer of fluid. This fluid, however, is not pure water. Water, as you may have noticed, has a tendency to form spherical drops when placed on a flat surface. The force holding up the drop against the pull of gravity is called **surface tension**. The hydrogen bonds (Chapter 4) that hold the water molecules together are stronger near the surface of the drop, as Figure 6.6 shows.

If the alveoli had only a water film inside, then the surface tension would try to roll up an alveolus like a drop and close it off. It would then take a lot more pressure to force it open and let the inspired air in. The water therefore is mixed with a substance called a **surfactant** and the particular one used in the lungs is a lipoprotein. The lipo part of the molecule is insoluble in water, so it forms a

Surface tension

A property of liquids. The surface appears to be covered with a thin elastic membrane caused by unbalanced cohesive forces among the molecules at the surface. Surface tension pulls water into spherical drops

Surfactant

A substance that can reduce the surface tension of its solvent, or the tension formed between one surface (such as water) and another (such as the walls of an alveolus).

FIGURE 6.6 Molecular theory of surface tension, showing attractive forces (only) on a molecule at the surface, and on one deep inside the liquid

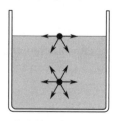

Source: *Physics: Principles with Applications*, 3rd Ed, Giancoli, D.C., 1991, p. 262. © Reprinted by permission of Pearson Education Inc., Upper Saddle River, N.J.

thin layer between the water and the inspired air. This reduces the surface tension, allowing the alveoli to remain open. Without it, breathing would be almost impossible. In fact, it reduces the pressure required to keep the alveoli open from a positive pressure of 20–30 kPa to only 3–5 kPa. In some cases, premature babies (less than about 28 weeks old) are born without this surfactant and some of them die shortly after.

There are approximately 300 million alveoli in the lungs, each of which is only 300 μm across. This is enough to give a total area of alveoli of close to 75 m². Since this area is lined with surfactant-filled water, there is plenty of fluid for the gases to dissolve in. In the case of Mr W., however, the progress of his emphysema has meant the destruction of many of his alveoli. The walls between them have broken down, forming fewer but larger alveoli. This means there is less surface area for the gases to dissolve in and so less is available for entry into his body. In the case study, we referred to this as a problem with gas exchange.

Diffusion of the gases into the bloodstream

This is the time to introduce one of the most important processes that occur in the body: **diffusion**. It involves the overall movement of particles, such as molecules, from a place where they are concentrated together, to a place where they are less concentrated. This movement happens naturally, without anyone or anything needing to do any work on the particles. As long as there are no barriers in the way, or the molecules are not tightly bonded into a solid, they'll move; Figure 6.7 demonstrates the diffusion of ink in water.

The reason why it happens is quite simple. All the molecules of a fluid (such as a gas or a liquid) are moving very fast and in a random way. They bounce off one another all the time and constantly change direction. It is important to keep in mind this idea of random motion, because it distinguishes diffusion from other movements, such as convection currents, where all the molecules move in roughly the same direction, which is clearly not random. Imagine there is an open space (area of low concentration) to one side of a big crowd of molecules (area of high concentration). As time goes by, some molecules in the crowd happen to shoot out into the open space. The crowd gets smaller and the open space fills up. After a while the original crowd is gone and the entire area is filled with molecules fairly evenly spread out.

To be more precise, however, we need to say a little more about diffusion than that and in particular about molecules in a gas, since we are focusing on their movements in our attempt to understand Mr W.'s COPD.

Diffusion
The movement of particles from an area of higher concentration to an area of lower concentration. It may or may not occur through a permeable membrane. Diffusion is passive (requires no outside energy) and results from the random motion of the particles

FIGURE 6.7 Diffusion.

Step 1: Placing an ink drop into a glass of water establishes a strong concentration gradient because there are many ink molecules in one location and none elsewhere. Step 2: Diffusion occurs, and the ink molecules spread through the solution. Step 3: Eventually diffusion eliminates the concentration gradient, and the ink molecules are distributed evenly.

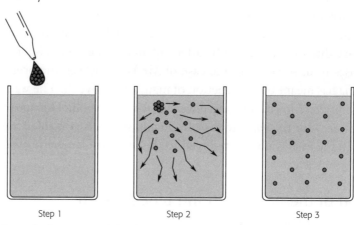

| Step 1 | Step 2 | Step 3 |

Source: *Fundamentals of Anatomy and Physiology*, 2nd Ed, Martini, F., 1992, p. 79. © Reprinted by permission of Pearson Education Inc., Upper Saddle River, N.J.

First of all, what do we mean by the word 'concentrated'? Imagine we have a bottle of soda water, which contains carbon dioxide dissolved in water. Of course, in this case, a 'concentrated soda water' means a lot of CO_2 dissolved in only a little water. We can talk about soda water being anywhere from very bubbly to flat or, technically, from concentrated to dilute.

This combination of a gas dissolved in water is one example of a **solution**. Whenever one substance is dissolved in another, we have a solution, which can be either dilute or concentrated. The substance that gets added to the liquid and dissolved is called the solute and the substance it gets dissolved into is called the solvent. That is,

solution = solute + solvent

The solute can be a solid, liquid or gas. The solvent can be either solid or liquid, though in the human body, water is by far the most common solvent. So as a simple example, soda water is a solution, with water as the solvent and carbon dioxide as the solute. In the alveoli, water again is the solvent and the gases that dissolve in it are the solutes. In the human body in particular, most

Solution

The combination of a solute (or solutes) dissolved in a solvent. The solutes are the substances (gases, liquids or solids) that are dissolved in the solvent, which is usually water.

solutions are a complex mixture of solutes dissolved in water. In this chapter we are concerned with solutes that are gases; in Chapter 7 we look more closely at solid and liquid solutes.

How is concentration measured? Quite simply, it's the ratio of solute to solvent. That is,

concentration = solute/solvent

Looking at our earlier example from Table 6.3, the concentration of O_2 dissolved in the watery film in the alveoli was given as:

3.3×10^{-6} mL/mL H_2O

As another example, under resting conditions we breathe in 4.2 L (4200 mL) of air per minute and O_2 is about 21% of that, or 0.88 litres (880 mL). This is a concentration of:

880 mL/4200 mL or 0.21 mL/mL of air = 21%

We use 250 mL of that O_2 in cellular respiration, leaving 630 mL. The new concentration is, therefore,

630 mL/4200 mL or 0.15 mL/mL of air = 15%

The amounts of solute and/or solvent do not have to be measured in mL or litres. They can also be measured in grams, for example. A very common way of measuring amounts of substances such as solutes uses the SI unit of amount, the mole. The mole and solutions are discussed fully in Chapter 7. The important point here is that we can calculate concentrations and thus find where the solute is more concentrated. When we look at an alveolus we see that the gases have very different concentrations indeed, depending on where we look. Figure 6.8 is a diagram of an alveolus with its surrounding membranes.

The membranes separating the dissolved gases in the alveolus and the blood in the capillaries are so thin (about 0.5 μm) that the gas molecules can fit through them. Strictly speaking, the molecules are passing through small openings called membrane channels, or pores, in the membrane. The membranes are said to be permeable to these gases. **Permeable** is from an old Latin word meaning 'to pass', or 'go through'. If you look back to the size of an atom given

Permeable (or semi-permeable)
The property of a membrane that allows the diffusion of certain particles to occur. Permeability refers to the presence of 'pores' in the membrane through which molecules of the right size, shape and electric charge can pass.

FIGURE 6.8 Alveolar organisation.

Connective tissue layers and vascular supply to the alveoli. A network of capillaries is surrounded by elastic fibres. respiratory bronchioles also contain wrappings of smooth muscle that can vary the diameter of those airways.

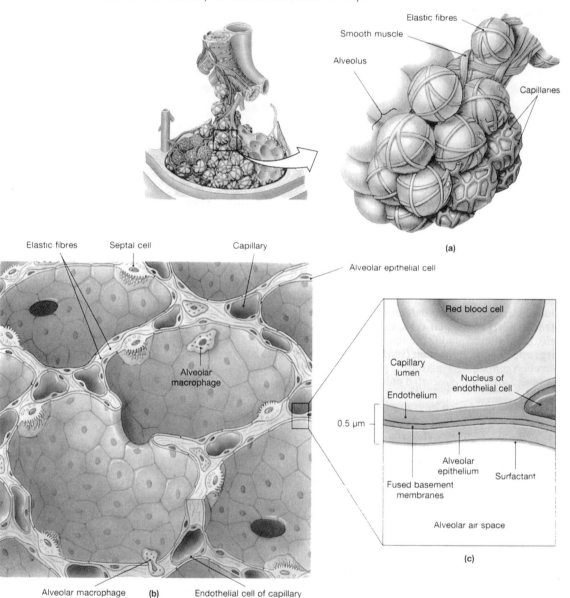

in Chapter 4, you'll see why a gas molecule is small enough to fit through the holes in something as 'solid' as a capillary membrane. Of course, if the molecule is too big to fit through the pores, as is often the case with large protein molecules, diffusion will not take place. When a membrane allows only molecules of a certain size to pass through its pores, it is said to be semi-permeable.

It is both convenient and traditional to use the partial pressure values for the concentration of the gases in the alveoli and the capillaries. This works out fine, because the amount of gas that is dissolved depends on the partial pressure anyway (that was Henry's Law, remember). Because we take in oxygen from the atmosphere when we inhale, the concentration of O_2 is obviously higher inside the alveolus than in the capillary, and the concentration of CO_2 is lower inside the alveolus than in the capillary; see Table 6.2.

Now diffusion can come into action. The O_2 moves from a region of high concentration to low and therefore enters the bloodstream. The CO_2 moves out from the blood and into the alveolus, to be expired. As long as we know the partial pressures, we can always predict where the gases will move from and to. The CO_2 now in the alveoli, at a partial pressure of 40 mm Hg, will flow out into the atmosphere, with a partial pressure of only 0.3 mm Hg, during expiration.

Factors affecting diffusion

The diffusion of all these gases is influenced by three main factors:

1. *The thickness of the membrane.* A thickness of 0.5 μm presents no real problems to the O_2 or CO_2, but if the membrane thickens there may be too much resistance to the smooth flow of molecules across the membrane. There is no evidence that this is a problem for Mr W. However, clients with pneumonia have extra fluid in the alveoli, which makes the distance the molecules have to travel greater.

2. *The surface area of the membrane*—the bigger, the better. Remember, the clinical supervisor's case notes mentioned Mr W.'s impaired O_2/CO_2 exchange. This is because many of his alveoli have joined together. These new alveoli are larger than the old ones, but the total surface area has been reduced. This reduction is sometimes by as much as two-thirds to three-quarters. It's like reducing the number of entry gates to a cricket match from 20 to five. The movement of people into the stadium is much reduced. The word 'emphysema' literally means 'too much air' and gets its name from

those enlarged alveoli. Mr W. can get the air into his lungs, he just can't get it into the blood.

3. *The partial pressure difference from one side of the membrane to another.* The greater the pressure difference, the greater the diffusion. The pressure is like a mechanical push, forcing the molecules through the small pores in the membrane. Again, this is not a particularly severe problem for Mr W.

Diffusion coefficient

A measure of the ease with which a particular molecule will diffuse through a particular membrane; the higher the coefficient, the greater the amount of diffusion.

To add a final detail at this point, diffusion is also influenced by the **diffusion coefficient**. Like the solubility coefficients listed in Table 6.3, it is specific for each molecule. It takes into account two things about the gas: its solubility in the membrane (not in the solvent) and its weight. Since weight is related to size, it's easy to see that weight would be important—the molecules must be small enough to fit through the pores. Remember, too, that cell membranes are made of lipoproteins and lipids are not soluble in water. The gases must be soluble in the lipid part of the membrane or they won't have as easy a ride through; see Figure 6.9.

Transport of gases by the blood

Once across the capillary membrane and into the blood, the O_2 is inside the body at last. However, only about 3% of that O_2 can dissolve in the blood since it is not very soluble, as shown by its low solubility coefficient. Since that 3% isn't nearly enough to meet the needs of cells, the rest has to be carried by other means. The red blood cells contain a molecule rich in iron called haemoglobin. The haemoglobin has a strong attraction for oxygen:

$$Hb + O_2 \rightleftarrows HbO_2$$

There are two things to say about this reaction: first, Hb is not the symbol of an element, but a shorthand way of writing haemoglobin; second, each haemoglobin molecule contains four heme groups, which contain the iron atoms, so each haemoglobin can in fact carry four O_2 molecules.

At this point we can recall some of the chemistry from Chapter 4. The reaction shown above has arrows going in both directions, which shows that the reaction can go in either direction; it is reversible. That is, we can break down an HbO_2 molecule into Hb and O_2 again. The arrows are the same length, which shows that both reactions happen at the same rate. Therefore, this is another example of an equilibrium reaction. But it's possible to upset that

FIGURE 6.9 Diffusion through cell membranes.

Small ions and soluble molecules diffuse through membrane channels. Lipid-soluble molecules can cross the membrane by diffusing through the phospholipid bilayer. Larger molecules that are not lipid-soluble cannot diffuse through the membrane at all.

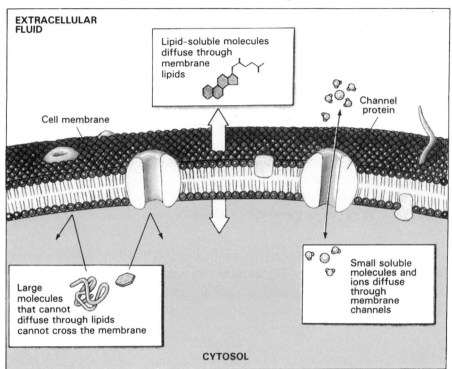

Source: *Fundamentals of Anatomy and Physiology*, 2nd Ed, Martini, F., 1992, p. 80. © Reprinted by permission of Pearson Education Inc., Upper Saddle River, N.J.

equilibrium. If we add a lot of O_2 to the blood, HbO_2 forms faster than it breaks down, as shown:

$$Hb + O_2 \rightleftharpoons HbO_2$$

If, on the other hand, O_2 is in low supply, the breakdown of HbO_2 takes place faster than its formation:

$$Hb + O_2 \rightleftharpoons HbO_2$$

Where is the amount of O_2 the greatest? Near the lungs, in the capillaries just outside the alveoli, of course. So it's there that HbO_2 forms fastest. Where is O_2 in shortest supply? Deep in the body, far from the lungs. It's not mere coincidence, then, that it's precisely there that the HbO_2 breaks down, releasing the needed O_2; see Figure 6.10.

This process is also under the influence of factors such as the pH of the blood and the concentration of carbon dioxide. More will be said about the transport of gases in the bloodstream in Chapter 8.

The CO_2 is generated by body cells and moves by diffusion from the cell into the blood. It needs to be transported to the lungs. Recall that CO_2 has a higher solubility coefficient than oxygen. As a result, about 7% of the CO_2 is dissolved in the blood. Another 23% is carried by haemoglobin which is free of oxygen. The remaining 70% is transported in the form of an important ion, bicarbonate ion. The way this happens is as follows.

When we add CO_2 to water, some of it dissolves, but a great deal of it reacts with water in a chemical reaction.

$$CO_2 + H_2O \rightleftharpoons H_2CO_3 \rightleftharpoons H^+ + HCO_3^-$$

The CO_2 from the cells is carried to the lungs as bicarbonate ion, where the above reaction goes into reverse, forming CO_2 molecules again; see Figure 6.11.

This process is discussed in greater detail in Chapters 7 and 8.

FIGURE 6.10 An overview of respiration and respiratory processes

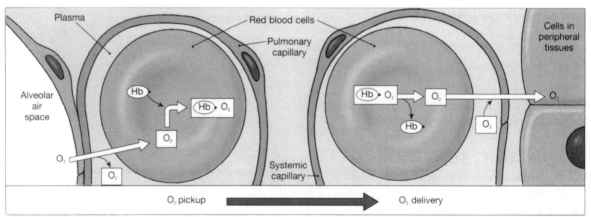

FIGURE 6.11 An overview of respiration and respiratory processes

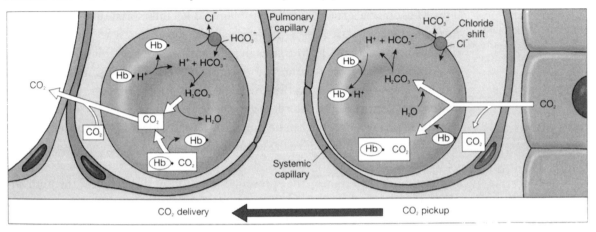

Source: *Fundamentals of Anatomy and Physiology*, 5th Ed, Martini et al, 2001, p. 831. © Reprinted by permission of Pearson Education Inc., Upper Saddle River, N.J.

Summary

1. Diffusion is the movement of solutes from areas where they are concentrated to areas where they are less concentrated, due to random molecular motion.
2. Diffusion may occur as the solutes move through a permeable membrane.
3. Some factors that affect this diffusion include size of molecule and of membrane pore, pressure differences, area and thickness of membrane and coefficient of diffusion.
4. Equilibrium reactions can be affected by changing the concentration of the reactants or products.

6.4 AIR POLLUTION

Open to the atmosphere at one end, the lungs are especially prone to harmful influences from the environment. The air is not particularly clean, even in areas remote from large urban centres or factories. With every breath we take, dust, spores, insect eggs, fumes and aerosols may enter the body. All of these can cause illness. Pathophysiologists usually group these together according to the type of damage they do, so we'll do the same.

Interstitial lung disorders (pneumoconiosis)

This is a diverse group of lung problems. The interstitial area of the lungs involves the connective tissue that gives shape and firmness to the alveoli. If it hardens up or gets too fibrous, the alveolar membrane gets thicker and denser, and diffusion therefore becomes more difficult. Also, with loss of elastic tissue, the lungs lose compliance. Breathing becomes more difficult. Some of the causes of interstitial lung disorders appear in Table 6.4.

As you can see, damage can be caused by a very wide range of substances and it would be impossible to give details of them all here. A few generalisations can be made, however.

The body reacts to the presence of 'foreign' particles through its immunological system. As the major body defence against invaders, it is very important and very complex. The details are beyond the scope of this book, but a good physiology text will help. The point here is that the production of antibodies to deal with the foreign particles may produce an allergy and what is called 'hypersensitivity'. For example, you may have heard of a group of chemicals called histamines. These are produced by certain cells in the body to fight infection. Histamines cause muscle contraction, too, and if this occurs in the lungs such contraction may be fatal. So the foreign particles may be harmful by causing the body's own defences to overreact.

Whether the particles themselves do any damage or not seems to depend on

TABLE 6.4 INTERSTITIAL LUNG DISORDERS

Causes	Examples
Inorganic dusts	Silicosis
	Asbestosis
	Talcosis
	Coal miners' pneumoconiosis
Organic dusts	Farmers' lung
	Pigeon breeders' lung
	Air-conditioner lung
Infections	Tuberculosis
Poisons	Paraquat (herbicide)
Disease	Chronic pulmonary oedema

four factors: (1) the concentration of the particles; (2) their size and shape; (3) their chemical composition; and (4) how long they remain in the lungs. Let's look at these one at a time.

Concentration of particles

The lungs are remarkably efficient at filtering out and removing most harmful and harmless substances. Any disease-causing particles therefore usually have to be present in large amounts. When discussing pollutants, it's common to talk about their concentration in **parts per million (ppm)** or **parts per billion (ppb)**; that is, one particle of pollutant per 1 million or 1 billion particles of air (or water, if we talk about water pollution). The following conversions may be helpful:

$$1 \text{ ppm} = 1 \text{ μg/g} = 1 \text{ mg/kg} \qquad 1 \text{ ppb} = 1 \text{ μg/kg}$$

For example, some people are affected by sulphur dioxide (SO_2) in smog at concentrations as low as 0.002 ppm. However, concentrations are not always measured so exactly. Farmers' lung disorder, for example, can occur whenever enough spores of the mouldy hay fungus reach the alveoli to irritate them. The irritation (called alveolitis) produces excess fluid in the alveoli, thus slowing diffusion. It also damages cells and this leads to scar tissue. The result is reduced pulmonary volumes and impaired gas exchange.

Parts per million (ppm) and **parts per billion (ppb)**
A measure of concentration, usually used in discussing pollutants. Technically, 1 ppm = 1 mg/kg
1 ppb = 1 μg/kg.

The size and shape of the particles

These two factors are important because they affect how well, or poorly, the particles are eaten by the lungs' macrophages, the wandering cells that locate and ingest foreign particles. If a large number of them are killed by the toxic particles they eat, the area of the body where this occurs is disturbed. Fluids are released by the macrophages and surrounding tissue, the alveoli become blocked with dead cells and the whole area is likely to become inflamed. In the case of asbestosis, for example, the most harmful particles are in the 20–100 μm size range. These become stuck in the bronchioles and the macrophages come after them. The asbestos fibres are toxic to these large cells, which die. The end result is fibrosis, or formation of tough, fibrous tissue around the bronchiole. This scar tissue is inelastic and obstructs the airways.

The chemical nature of the particles

While some are harmless, some interfere directly in the body's chemical homeostasis. Cadmium metal poisoning is a good example of this. Once inspired, cadmium enters the cells of the body, not by diffusion but by being carried in by a protein molecule. Once inside, the cadmium sticks onto the membranes that line the mitochondria. There it interferes with both the Krebs cycle and the electron transport chain, discussed in Chapter 5. The production of ATP is cut down and the cell dies from lack of energy.

The duration of exposure

Duration of exposure is important because damage is seldom detectable until the number of particles has accumulated to such a level that the natural cleaning of the lungs can no longer cope. If exposure to toxic chemicals in the air happens infrequently, the lungs can usually flush them out. However, the natural action of cilia in the respiratory tree, which sweep out unwanted particles, can become overpowered by long exposure to these contaminants.

Chemical injury (poisoning)

Let's concentrate here on gases, which can use only the lungs as entry points into the body. The classic example is carbon monoxide poisoning. Carbon monoxide (CO) has a great affinity for haemoglobin. Once in the blood via diffusion, it sticks to the haemoglobin, which can no longer take on oxygen. To add to the problem, it sticks so tightly that it is only very slowly removed from the body. The exhaust fumes from cars contain about 7% carbon monoxide. In a closed garage, fatal levels of CO can be reached in 5 minutes.

Other gases present in the atmosphere that cause damage are those that form when sulphur (S) is released and when nitrogen (N) is heated to very high temperatures. Let's look at the sulphur compounds first. These are mainly sulphur dioxide (SO_2) and hydrogen sulfide (H_2S). When SO_2 dissolves in water, it becomes sulphuric acid (H_2SO_4). If this occurs with water in the atmosphere, we get 'acid rain'. If the water is in the lungs, we get upper respiratory tract irritation, at a concentration of 1 ppm. At doses of 36 ppm, there is a high degree of coughing and dyspnoea (difficulty in breathing).

Nitrogen is normally harmless. When strongly heated, as in car engines, it can produce nitrous oxide (NO), which can go on to form nitrogen dioxide (NO_2). The exact way NO_2 causes damage is uncertain. It seems to cause damage (at levels of only 1 ppm) to the elastic fibres of the lungs and may therefore be linked to Mr W.'s emphysema. This damage also leads to the release of histamine, mentioned above. It is also involved in interference with the Krebs cycle.

Lung tumours

Cancers of the pulmonary system are mentioned here only to highlight the relationship between the inspiration of tobacco smoke (an environmental gas) and lung disease. Table 6.5 gives some idea of the types of lung tumours and how common they are.

The interesting thing is that bronchogenic carcinoma is uncommon in non-smokers. There are several hypotheses about how tobacco smoke brings about this carcinoma. One suggests that the smoke particles are catabolised into chemicals that directly interfere with the DNA of a cell. A second view is that the particles act as a constant, low-level irritant. The exact connection between this irritation and the outbreak of cancer is still unclear.

Finally, it should be mentioned that smoking also has strong links to emphysema. Mr W. was a heavy smoker for many years and samples of his lung tissue show the patterns we have discussed throughout this chapter: the joining of small alveoli into larger ones, the loss of elasticity through scarring and the accumulation of fluid (oedema) in the alveoli due to lots of dead macrophages (the infection-fighting cells). The long-term outlook for Mr W. is not good, as improvement in lungs so damaged is not likely to occur at his age.

TABLE 6.5 TYPES OF LUNG TUMOUR

Type	Per cent
Bronchogenic carcinoma	90
Alveolar cell carcinoma	2
Bronchial adenoma	5
Mesenchymal tumours	1.4
Miscellaneous	1.5

Questions

Level 1

1. Define the following and give an example of each:

   Boyle's Law

   Charles' Law

   Henry's Law

   Dalton's Law

2. Use the general gas law to find the volume of air exhaled under the following conditions:

   $P_1 = 98$ kPa $P_2 = 105$ kPa

   $T_1 = 25°C$ $T_2 = 35°C$

   $V_1 = 500$ mL $V_2 = ?$

3. Define ventilation. Explain how air moves into and out of the lungs, using the concept of pressure.

4. What is meant by negative pressure in the lungs? How does intrapleural pressure prevent the lungs from collapsing?

5. What is the importance of the residual volume?

6. What is diffusion? Use your definition to account for the movement of gases to and from the blood through the walls of the alveoli and capillaries. What factors will affect this process?

7. What is the partial pressure of a gas? How does it help to explain how gases move in the alveoli and surrounding capillaries?

8. For the following solutions, identify solute and solvent:

   (a) a beaker of salt water

   (b) a 10% dextrose solution for IV therapy

   (c) 5 g of butter dissolved in a litre of peanut oil

   (d) 3×10^{-6} mL of O_2 dissolved in 1 mL of water

Level 2

9. Why does emphysema affect the surface area of the alveoli and therefore the amount of oxygen entering the body?

10. What effect would Mr W.'s emphysema have on his expiratory flow rate?

11. Explain the process by which oxygen is transported in the blood and released to the cells in the body.

12. Why is the presence of a surfactant in the fluids of the alveoli important?

13. How is concentration measured? What is meant by a 0.9% NaCl solution?

Level 3

14. A person's vital capacity is measured while they are standing and again while they are lying down. What difference in the measurement would you expect and why?
15. Find out what smog is. What are some of its effects on the body?
16. Another factor that affects the process of diffusion is the presence of electric charges due to ions on one or both sides of the cell membrane. Discuss how this might either increase or decrease the rate of diffusion.
17. Mr W.'s case notes refer to the delivery of low oxygen concentrations to avoid the loss of 'hypoxic respiratory drive'. What does this mean?

CALCULATION PROBLEM BASED ON THE CASE STUDY

Mr W. is being treated with an IV bronchodilator, theophylline. He is ordered a mainte-nance dose of theophylline 375 mg in 500 mL of NaCl 0.9% by slow IV infusion to be given over 6 hours. The drop factor is 20. On hand is 250 mg theophylline in 10 mL ampoules. Calculate the injection volume to be added to the saline drip, and the speed of the drip to run over 6 hours.

Answer:
Injection volume = 375 mg/250 mg × 10 mL = 15 mL theophylline.
Rate (drops/min) = 500 mL/6 hr × 20 drops/mL/60 min/hr = 28 drops/min.

7

THE RENAL UNIT AND CLIENT CARE

MORE ON DIFFUSION, OSMOSIS AND AN INTRODUCTION TO ACIDS

Chapter outline

INTRODUCTION

The key theme running through Chapters 6, 7 and 8 is that of the nature and properties of solutions. In Chapter 6 we looked at how specific gas solutes, O_2 and CO_2, move between the atmosphere and the bloodstream to meet the needs of cells deep within the body. In this chapter we continue that discussion, to include both other important solutes, such as acids and bases, and other methods by which they enter cells. But we also turn our attention to the solvent part of solutions. We need to understand the way water can move in or out of cells and the bloodstream. And finally, we look more closely at how solution concentrations are measured.

Case Study: Mrs Verna B.

Mrs B. is an alert, intelligent woman of 35 years. She has been using a dialysis machine for the last 5 years, ever since contracting acute glomerulonephritis. She is quite capable of describing, in her own words, what the machine is doing for her.

'This machine cleans my blood. The blood flows out from my artery here,' she says, pointing to her left arm, 'and goes into the machine. It then passes over a filter with very tiny holes. The waste products pass through the holes, leaving the blood clean. All the time, the blood is kept at my body temperature, then it's pumped back into my vein here.'

The CNC of this unit informs you that Mrs B. is on a short list for a kidney transplant. Until then she is treated with drug therapy, diet and the dialysis machine, for oedema, acidosis and high urea levels (uraemia).

Clinical notes: The expected outcomes for Verna B. include adequate fluid balance, as evidenced by blood pressure within normal range, a heart rate less than 100 bpm, a cardiac output of 4–8 L/min, and haemoglobin and haematocrit within normal range.

7.1 THE MOVEMENT OF MOLECULES IN AND OUT OF CELLS

We said in Chapter 6 that diffusion is one of the most important means by which nurses understand how molecules and ions move across cell membranes. Let's have another look at it here, remembering that it refers to the random movement of particles from places of higher concentration to places of lower concentration. The particles discussed in Chapter 6 were gas molecules dissolved in water. The gas molecules were the solute, the water was the solvent and the combination was the solution.

It will be helpful to begin with a simple distinction. The fluids of the body can be divided into those that occur in the cells, called intracellular fluid (ICF), and those that are outside the cells, called extracellular fluid (ECF); see Table 7.1.

In both the ICF and ECF the solvent is again water. The number and type of solutes in both, however, is very large. There are molecules of gases such as CO_2 and O_2 and N_2, for example. There are large protein molecules, glucose and other sugars, ions such as chloride (Cl^-), sodium (Na^+) and potassium (K^+) and many other simple and complex molecules. Some of these solutes are present in almost equal amounts in the ICF and ECF and some occur mainly in one or the other. As well, some solutes appear only now and again (such as after a meal), or only after the person reaches a certain age (like the sex hormones). Finally, keep in mind also the varying roles these solutes play: some are nutrients, some are wastes, some are messengers, some are building materials.

A solute is normally dissolved in a solvent. Figure 7.1 shows this schematically.

TABLE 7.1 A COMPARISON OF ECF AND ICF FLUIDS

Component	ICF	ECF
Ions (mmol/L)		
Na+	10	145
K+	160	4
Cl-	3	114
Dissolved protein (g/100 mL)	16	1.8
Nutrients (mg/100 mL)		
carbohydrates	0–20	90
amino acids	200	30
lipids (g/100 mL)	2–95	0.6

FIGURE 7.1 The dissolving of NaCl by water.

Polar water molecules attract Na+ and Cl−; the ions then break away from the solute and move into solution, where they are hydrated.

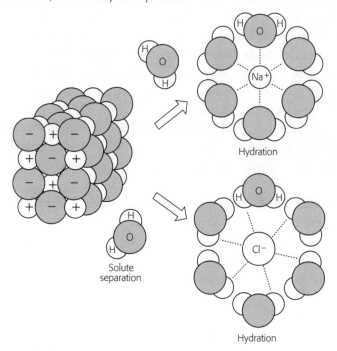

Hydration

Solute
separation

Hydration

Source: *Chemistry*, 5th Ed, Timberlake, K.C., 1992, HarperCollins, New York, p. 274.

The water molecules split the solute molecule into smaller pieces and form weak hydrogen bonds with them. However, in some cases the solute particles do not dissolve, but are too small (about 1–1000 nm) to sink to the bottom of the solvent; the solution is called a **colloid**. A common example is the proteins in the blood. Proteins are fairly large molecules, but are still too small to settle to the bottom, or be seen by the naked eye. The solute particles in a colloid scatter light, making their presence known; this is the Tyndall effect; see Figure 7.2.

In a colloid the water molecules do not bond with the solute particles, but bounce against them so often and so hard they never sink to the bottom.

A third possibility is when the solute doesn't dissolve but does sink to the bottom. This is referred to as a **suspension**. Medicines with labels that ask you to shake before administering are often suspensions and blood itself contains many substances (such as the red blood cells) that settle to the bottom if the blood is left standing.

Figure 7.3 shows a model of a typical cell with sodium ions as our

Colloid

A solution in which the solute particles are not dissolved, yet are too small to sink to the bottom.

Suspension

A solution in which the solute particles are not dissolved and are large enough to settle to the bottom over time.

FIGURE 7.2 The Tyndall effect

Light is reflected by colloid particles, making a light beam visible; true solution particles are too small and do not produce a visible beam. This is known as the Tyndall effect.

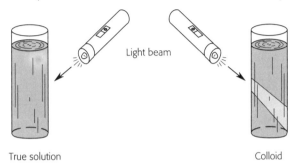

True solution   Colloid

Source: *Chemistry*, 4th Ed, Timberlake, K.C., 1988, Harper and Row, New York, p. 263.

FIGURE 7.3 Model of the process of diffusion of sodium ions through a semi-permeable membrane

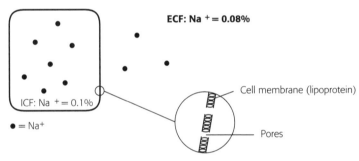

representative solute. For the purposes of this example, assume that the ICF has a Na^+ concentration of 0.1% (0.1 g of Na^+ in 100 mL of water) and the ECF value is 0.08%.

Separating the ICF and the ECF is the cell membrane. Recall from Chapter 6 that the membrane contains numerous membrane channels, or pores, which account for its permeability. Since these pores are large enough for Na^+ ions to pass through, this is an example of passive diffusion. The cell does not use energy to assist or hinder the natural diffusion across its membrane. Na^+ ions will move into the cell until the concentration of the ICF equals that of the ECF.

Passive diffusion is not, as you might have guessed, the only kind of diffusion. There are some molecules that a cell might be particularly interested in having, but it wants greater control over how many enter and for this purpose uses a process called **facilitated diffusion**; see Figure 7.4.

Facilitated diffusion
Diffusion in which the solute particles are assisted through the cell membrane by carrier molecules designed for that purpose.

FIGURE 7.4 A model of facilitated diffusion, where a molecule is transported across the membrane by a carrier protein

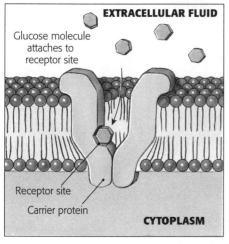

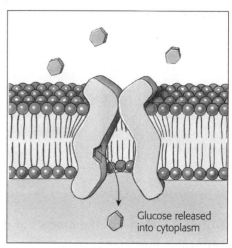

Source: *Fundamentals of Anatomy and Physiology*, 5th Ed, Martini et al, 2001, p. 75. © Reprinted by permission of Pearson Education Inc., Upper Saddle River, N.J.

There is a higher concentration of glucose molecules outside the cell than inside. But in this case, there is a special 'carrier' molecule in the membrane of the cell that is just the right shape to latch onto glucose. Once it has got hold of a glucose molecule, it passes it through the membrane and releases it in the interior. This process still only allows movement from high concentration of solute to low, however. The only difference is the presence of the carrier molecules. If they are not present, the molecule cannot enter the cell.

A third possibility is referred to as **active transport**. This allows the cell, for example, to take molecules from an ECF of low solute concentration and move them into an ICF of higher solute concentration; see Figure 7.5.

This is, of course, directly opposite to diffusion and takes energy for the cell to accomplish. It allows the cell to concentrate desired molecules or ions, such as K^+ and Mg^{2+}, even if they are in short supply in the surrounding bloodstream.

To relate all this to Mrs B., consider just one aspect of her loss of kidney function. Throughout her body, many catabolic reactions are occurring as a normal part of cell metabolism. The breakdown of some complex molecules releases potassium ions, K^+, which pass by passive diffusion into the ECF. The kidneys are normally responsible for removing this K^+. Therefore, with a decrease in kidney function, several problems arise. Since there is less blood being filtered there is a greater danger of substances such as urea (from protein breakdown), acids (produced by metabolism, and discussed later) and water to

Active transport

The movement of solute particles across a cell membrane, using the cell's energy. It may move particles from low to high concentrations.

FIGURE 7.5 A model of active transport, showing the hingelike motion of the integral protein subunits

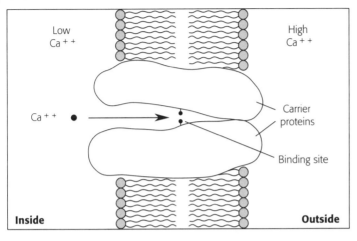

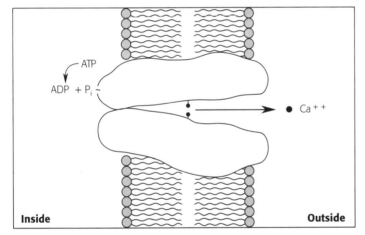

Source: *Concepts of Human Anatomy and Physiology,* 2nd Ed, Van de Graaf, K.M. and Fox, S.I., 1989, Wm. C. Brown, Iowa, p. 121.

accumulate in body tissues. In Mrs B.'s case especially, there may be an excess of K^+ building up in the ECF. It has a cardiotoxic effect, which leads to irregular heartbeat, cardiac failure and ultimately death.

Osmosis

So far we've talked about the solute part of a solution, moving across membranes by passive or facilitated diffusion, or by active transport. You may have wondered about the solvent part; can it move as well? The answer is yes

Osmosis
The movement of solvent molecules (usually water) from areas of higher solvent concentration to lower solvent concentration. The movement is passive and occurs across a membrane permeable to the solvent.

and the name for this movement is **osmosis**. It comes from a Greek word meaning the 'process of pushing' and indeed the water molecules can be thought of as being pushed from a high solvent concentration into a lower solvent concentration. This means that the water is moving from a dilute solution, where there is a low concentration of solutes, into a more concentrated solution. Osmosis seems to make sense, but take the time to think carefully about what is actually occurring. Consider Figure 7.6, depicting a cell surrounded by interstitial fluid.

We are going to look at the movements of solutes and solvents here, using the concentrations from Table 7.1

Look first at the ICF. Here there are two solutes shown: Na^+ at a concentration of 10 mmol/L, and proteins at 16 g/100 mL. Now look at the ECF. Here there are the same two solutes shown: Na^+ at 145 mmol/L, and proteins at

FIGURE 7.6 Model of interstitial processes of diffusion and osmosis: a cell surrounded by interstitial fluid

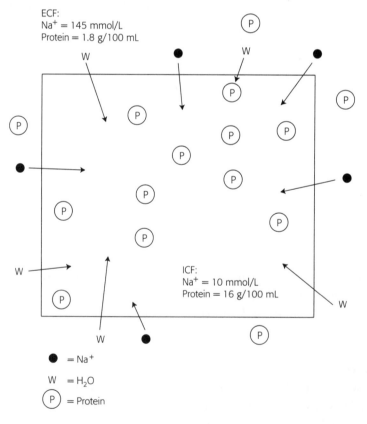

ECF:
Na^+ = 145 mmol/L
Protein = 1.8 g/100 mL

ICF:
Na^+ = 10 mmol/L
Protein = 16 g/100 mL

● = Na^+

W = H_2O

(P) = Protein

1.8 g/100 mL. These two solutions are separated by a semi-permeable membrane, which allows Na^+ to move through the membrane channels by diffusion, but does not allow the movement of protein.

As a result, the Na^+ ions will tend to move by passive diffusion from the ECF into the ICF, until the concentrations are equal, or until the cell takes active steps to either prevent diffusion (by closing the pores) or by pumping it back out again. The proteins, however, cannot move through the membrane and so remain at their present concentrations where they are.

You may be wondering why we bothered with the proteins in this case. They are actually crucial here for causing osmosis to occur. The requirements for osmosis are three. First, the two solutions must be separated by a semi-permeable membrane. The example above meets this requirement. Second, there must be a concentration difference between the two solutions. Again, our example meets this requirement. And third, there must be at least one solute present that is non-diffusible. Our proteins meet this requirement. Why is this important? Well, if all the solutes could freely move between the two solutions, the solutions would quickly end up with the same concentration, making osmosis impossible (this is our second requirement, remember).

Now, where is the concentration of solvent the highest? We can ignore the diffusing Na^+ here and concentrate on the proteins. Since there are many more protein molecules in the ICF, it will have less room for water molecules. So the ICF will have fewer water molecules per given volume of solution than the ECF will. Another way of saying this is that the ICF has a lower solvent concentration. In what directions will the water molecules go as a result? From the high solvent concentration ECF to the lower solvent concentration ICF. Water will enter the cell, causing it to swell.

One way to help remember this is as follows:

□ High solute concentration means low solvent concentration.
□ Low solute concentration means high solvent concentration.

Osmotic pressure
The solvent pressure that builds up as a result of the accumulation of solvent (say, in a cell) caused by osmosis. Osmotic pressure can be great enough to prevent further osmosis from occurring.

With all that water entering, the cell in our example would have to swell up like a balloon, due to the extra water. In fact, under extreme concentration differences a cell can actually burst, an unfortunate event known, remember, as lysis. If the cell doesn't reach the bursting stage, it might gain enough water pressure (referred to as hydrostatic pressure) to start pushing water back out as fast as it comes in. This solvent pressure pushing against the osmotic movement of water into the cell is referred to as **osmotic pressure**.

Technically, osmotic pressure is defined as the pressure that must be applied to a solution (such as the ICF), separated by a semi-permeable membrane from a solvent (such as the ECF), to prevent the movement of the solvent through the membrane. When the osmotic pressure pushes water out of a cell as fast as osmosis tries to push it in, equilibrium occurs.

Let's have a look at this in the kidney, using glomerular filtration pressure as an example, shown in Figure 7.7.

As the blood passes through the glomerulus in the nephron, it is filtered. Just as we use filter paper to keep the coffee grounds out of the percolator, the glomerulus contains tiny pores and slits which filter all but the largest molecules and cells in the blood (the larger plasma proteins and formed elements), thus leaving them behind.

The difference in pressure between the blood and the space in the glomerulus is about 10–15 mm Hg (about 2 kPa) when the blood enters the glomerulus and drops to 0 kPa at exit. This pressure difference is the resultant of three different pressures which act on the blood as it flows through a nephron. First is the blood pressure in the capillaries that surround the glomerulus. This is known as glomerular hydrostatic pressure (G_{hp}); it has a typical value of 60 mm Hg and forces water and solute particles across the glomerular wall, through the little slits.

The second pressure difference, referred to as capsular hydrostatic pressure

FIGURE 7.7 Glomerular filtration pressure.

A schematic diagram showing the size of three different pressures in relation to distance through the glomerulus.

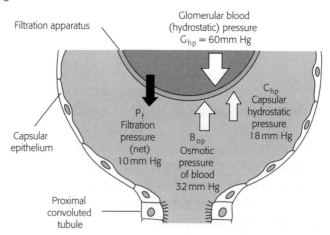

Source: *Fundamentals of Anatomy and Physiology*, 2nd Ed, Martini, F., 1992, p. 874. © Reprinted by permission of Pearson Education Inc., Upper Saddle River, N.J.

(C_{hp}), is caused by the water pressure of the solution just pushed into the capsule by the G_{hp}. It represents the resistance to the flow of solutions along the nephron and therefore works against the glomerular hydrostatic pressure. It has a value of about 18 mm Hg.

The third pressure difference results from the osmotic pressure of the blood (B_{op}), which also opposes glomerular hydrostatic pressure. It has a value of about 32 mm Hg. The blood osmotic pressure is caused by large protein molecules which form a colloid. Their high concentration in the blood in the capillary means high solute concentration and therefore low solvent concentration. Water tries to move into the capillary from the capsule by osmosis, resulting in a pressure in the capillary of between about 2.6 and 4.6 kPa (20–35 mm Hg).

If we add all of these pressures together, we have:

$$\text{filtration pressure} = G_{hp} - C_{hp} - B_{op}$$
$$= 60 \text{ mm Hg} - 18 \text{ mm Hg} - 32 \text{ mm Hg}$$
$$= 10 \text{ mm Hg}$$

This filtration pressure forces the fluids of the blood into the space in the renal capsule. The rate of filtration is about 105 mL of fluid per minute for healthy adult females and about 140 mL per minute for a healthy adult male.

Not all of the material that is filtered from the blood is unwanted. In fact, it's important to retain most of it for further use, so the nephron reabsorbs the desired nutrients and other molecules in the section called the convoluted tubules. The absorption depends very strongly on the number of carrier molecules present, both for facilitated diffusion and active transport. Unwanted wastes, toxins and excess molecular compounds are not reabsorbed and end up being excreted.

For Mrs B., the loss of kidney function means that her filtration rate is lower than normal, so excess water is retained by the body. This is a condition known as oedema. More specifically for Mrs B., this is both ICF oedema and ECF oedema, because five-eighths of the water she drinks enters the cells and three-eighths remains in the ECF. However, if her diet contains a high quantity of salt, the ECF then has a high solute, low solvent concentration and water moves out from the cell into the ECF, because of the greatly increased amounts of osmotically active substances there.

Table 7.2 summarises the various processes discussed above.

Understanding all this, we are now in a better position to describe what goes

TABLE 7.2 SOLUTE/SOLVENT TRANSPORT

Passive diffusion	The unassisted movement of solute particles from an area of high solute concentration to an area of lower solute concentration.
Facilitated diffusion	Carrier molecules, of the right shape, transport the solute molecules across the membrane, again from high concentration to low concentration.
Active transport	Carrier molecules use the cell's energy to transport solute molecules across the membrane, from low solute concentration to high concentration.
Filtration	Differences in hydrostatic and osmotic pressure force fluid through a filter and solutes are left behind.
Osmosis	The unassisted movement of solvent particles (usually water) through a semi-permeable membrane, from an area of high solvent concentration to an area of lower solvent concentration.

Dialysis
The use of diffusion and/or osmosis to remove unwanted solutes/solvents from a solution.

on in Mrs B.'s artificial kidney machine. The process of separating particles from fluids is known as **dialysis** and since the particles are being removed from the blood (*haem*, in Greek), the machine is sometimes called a haemodialysis machine. Dialysis takes advantage of diffusion and osmosis to remove unwanted chemicals and to add necessary ones. Figure 7.8 is a simplified sketch of the machine and its functions.

Mrs B.'s blood is circulated through a series of small tubules made from a semi-permeable membrane. The membrane is surrounded by a solution called dialysis fluid, or dialysate. Waste products in the blood are removed by passive diffusion and any extra water causing oedema is removed by osmosis. Valuable nutrients and ions that must be retained are kept from diffusing out of the blood by maintaining a higher concentration in the dialysis fluid. From Table 7.3 note how the concentrations allow the molecules to move in the desired directions.

Although it varies somewhat from client to client, haemodialysis can take up to 15 hours per week and in between times, of course, the symptoms of uraemia, or nitrogen waste products in the blood, gradually reappear.

FIGURE 7.8 Kidney dialysis.

During kidney dialysis, blood flows through a system of tubes composed of a selectively permeable membrane. Dialysis fluid, the composition of which is similar to that of blood, except that the concentration of waste products is very low, flows in the opposite direction on the outside of the dialysis tubes. Consequently, waste products such as urea diffuse from the blood into the dialysis fluid. Other substances such as sodium, potassium and glucose do not rapidly diffuse from the blood into the dialysis fluid because there is no concentration gradient, since these substances are present in the dialysis fluid.

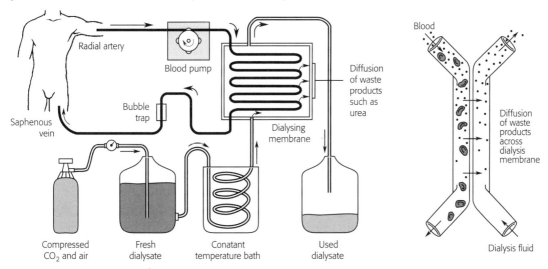

Source: *Anatomy and Physiology*, Seeley, R.R. et al., 1992, Mosby, St Louis, p. 860. Reproduced with the permission of the McGraw-Hill Companies.

TABLE 7.3 THE COMPOSITION OF DIALYSIS FLUID

Component	Normal blood (mmol/L)	Dialysis fluid (mmol/L)
Ions		
(K^+)	5	3
(HCO_3^-)	27	36
(HPO_4^{2-})	3	0
(SO_4^{2-})	0.5	0
Nutrients		
Glucose	100	125
Wastes		
Urea	26	0
Uric acid	0.3	0

Note: The units these substances are being measured in are discussed in Section 7.3.

7.2 ACIDS AND BASES

You may recall that in discussing Mrs B.'s condition the CNC mentioned **acidosis**. This refers to an excess of H⁺ ion in the body. This section of the chapter looks at how that comes about and how it is measured.

As we might expect, there is a link between an acid and the H⁺ ions. In fact, an **acid** is defined as a chemical that releases an H⁺ ion (a proton) when it dissolves in water, or reacts with other chemicals. For example:

$$\text{HCl (added to } H_2O) \rightarrow H^+ + Cl^-$$
hydrochloric acid

$$H_2CO_3 \text{ (added to } H_2O) \rightleftarrows H^+ + HCO_3^-$$
carbonic acid

$$CH_3COOH \text{ (added to } H_2O) \rightleftarrows H^+ + CH_3COO^-$$
acetic acid (vinegar)

Figure 7.9 puts this a slightly different way.

It is the H⁺ ion that makes the molecule an acid; the rest of the molecule determines what kind of acid it will be. The H⁺, rather than the rest of the molecule, is primarily responsible for the important health aspects of acids.

Bases, also referred to as **alkalis**, are chemicals that release an OH⁻ ion when they dissolve or react with other chemicals; and they also absorb H⁺. An OH⁻ ion is called an hydroxide ion. Here are a few examples:

$$\text{NaOH (when added to } H_2O) \rightarrow Na^+ + OH^-$$
sodium hydroxide

$$\text{KOH (when added to } H_2O) \rightarrow K^+ + OH^-$$
potassium hydroxide

FIGURE 7.9 H⁺ makes the molecule an acid; the different shapes indicate three different types of acid.

Acid A   Acid B   Acid C

Margin notes

Acidosis
The condition of excess H⁺ ions in the body.

Alkalosis
The condition of excess OH⁻ ions in the body, or lowered levels of H⁺ ions.

Acid
Molecules that release an H⁺ ion when they dissociate.

Base (or alkali)
A molecule that both accepts an H⁺ ion during a chemical reaction and releases an OH⁻ ion (hydroxide ion) when it dissociates.

$$NH_3 + H_2O \rightleftarrows NH_4^+ + OH^-$$
ammonia

The other part of the definition refers to a base as a chemical that can accept H^+ ions when they are around; Figure 7.10 illustrates this point.

In the example with ammonia (NH_3) above, the ammonia has taken an H^+ from a water molecule. Since acids supply H^+ ions and bases accept them, we can see that acids and bases react together by moving the H^+ around:

$$HCl + NaOH \rightarrow H^+ + Cl^- + Na^+ + OH^- \rightarrow NaCl + H_2O$$

In this case, the reaction produces common table salt (NaCl) and water. The acid and the base have effectively **neutralised** each other.

The information above is merely an example of the many types of reactions that acids and bases are involved in. Since the chemistry of these chemical compounds is so important in health and in industry, let's look at them in greater detail.

> **Acid/base neutralisation**
> The chemical reaction between an acid and a base, in which they form products (usually a salt and water) which are neither acidic nor basic.

Properties of acids

1. Acids release hydrogen ions when placed in water. This property we've examined above.
2. Acid solutions have a sour taste. This is obvious when we think of lemons (citric acid), vinegar (acetic acid) and sour milk (lactic acid).
3. Acids react with metals to release hydrogen gas and a salt. Examples are:

$$H_2SO_4 \quad + \quad Zn \quad \rightarrow \quad ZnSO_4 \quad + \quad H_2 \text{ (gas)}$$
sulphuric acid $\quad$ zinc $\qquad\qquad$ a salt

$$2HCl \quad + \quad Mg \quad \rightarrow \quad MgCl_2 \quad + \quad H_2 \text{ (gas)}$$
hydrochloric acid $\quad$ magnesium $\qquad$ a salt

FIGURE 7.10 Bases can accept an H⁺ ion from an acid; the different shapes indicate three different types of base.

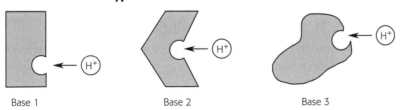

Base 1 $\qquad\qquad\qquad$ Base 2 $\qquad\qquad\qquad$ Base 3

4. Acids react with substances called carbonates and bicarbonates to form carbon dioxide, water and salts:

2HCl	+	$CaCO_3$	$\rightarrow$	$CaCl_2$	+	H_2CO_3
hydrochloric acid		calcium carbonate		salt		carbonic acid

then, $H_2CO_3 \rightleftarrows CO_2 + H_2O$

5. Strong acids destroy organic fibres such as cotton, linen, wood, wool and silk, and many inorganic fibres such as nylon. This effect applies also to skin and other body tissues. The term for this is corrosion and it is due to the high reactivity of H^+ ions.

6. Acids are found in many places in the body; e.g. as HCl in the stomach, where it aids enzyme function and kills some foreign microorganisms; as amino acids which make up protein; as lactic acid, associated with energy use in the muscles; and as ascorbic acid, Vitamin C. In these cases, the acids are not corrosive either because they are well diluted, or because they are not strong acids; see the section below on acid/base strength.

Properties of bases

1. Bases are H^+ acceptors. This also has been discussed above.
2. Solutions of bases taste bitter. The bitter taste of soaps and detergents is due to their base content.
3. Strong bases react with some metals to make hydrogen gas:

6NaOH	+	2Al	+	$6H_2O$	$\rightarrow$	$3H_2$	+	$2Na_3Al(OH)_6$
sodium hydroxide		aluminium						sodium aluminate

4. Bases can cause burns to tissue because they react with fats and proteins. The term used for this is caustic. An example is caustic soda, also called lye, which is chemically sodium hydroxide (NaOH) and used for cleaning.

The strength of acid and base solutions

When an acid is added to water, an H^+ ion is released. An example we used earlier is hydrochloric acid:

$$HCl \rightarrow H^+ + Cl^-$$

Every molecule of HCl is split into the two ions when added to water. The technical term for this splitting is **dissociation** and we say that HCl is completely dissociated into its ions. The arrow in the reaction points in one direction only, from the reactant to the products. Because it dissociates completely, HCl releases all the H⁺ ions possible for it and so it is called a strong acid. All strong acids dissociate almost completely. A solution of strong acid contains only the ions and none of the acid molecules.

Carbonic acid, however, does not dissociate very much at all:

$$H_2CO_3 \rightleftharpoons H^+ + HCO_3^-$$

The arrows point in both directions, meaning that, yes, carbonic acid dissociates into its ions, but the ions combine together again to remake carbonic acid. Carbonic acid, therefore, does not release as much H⁺ as possible and it is a weak acid. A solution of weak acid contains molecules of the acid as well as the ions it dissociates into.

This same distinction is used for strong and weak bases. Strong bases dissociate completely into ions. Weak bases dissociate only a little into their ions. It makes sense, then, that if we add a strong acid to water, the solution will have a lot more H⁺ ions in it than if we had added a weak acid. The concentration of H⁺ in a litre of water is higher for a strong acid than for a weak acid in a litre of water. This concentration of H⁺ is given the special measure, pH. Here's how it works. Say we add some HCl to water and find that the concentration of H⁺ is 10^{-4} M. (M stands for molar solution, discussed in a moment). The pH of this solution is simply the opposite of the –4 exponent, or pH = 4. Mathematically, what we are doing is finding the negative logarithm of the H⁺ concentration, but practically we are simply finding the negative of the concentration exponent. For example:

□ if H⁺ concentration = 10^{-3} M, then pH = 3
□ if H⁺ concentration = 10^{-6} M, then pH = 6
□ if H⁺ concentration = 10^{-12} M, then pH = 12

It's important to think very carefully about what the pH is telling you. A concentration of 10^{-3} M is 1000 times greater than a concentration of 10^{-6} M. Therefore a pH of 3 means a H⁺ concentration 1000 times greater than a pH of 6.

The lower the pH number, the higher the H⁺ concentration.

Dissociation
The breakdown of a solute into ions or smaller molecular units.

It turns out that water, which is neither acid or base, has a pH of 7. So pH = 7 indicates a neutral solution. Acidic solutions have pH values from 0 to 7 and basic solutions have pH values from 7 to 14. Table 7.4 gives the pH values for some common substances.

It's time to return to Mrs B. Table 7.4 shows that her urine should have a pH of between 5.5 and 7.5. This means it ranges from being acidic to slightly basic, depending on diet. But Mrs B. is not producing any significant quantities of urine at this late stage of her illness, because of her damaged kidneys. As a result, her blood and intercellular fluid show a consistent pH reading of 7, which is why the CNC mentioned acidosis. Even though a pH of 7 is chemically neutral, acidosis refers to the fact that there is an excess of H^+ in these body fluids. This leads to two questions: why is this a problem and how does a healthy kidney normally handle it?

Acids, or better still the H^+ ions they produce, are crucial to correct body functioning. A great deal more will be said about acidosis and alkalosis in the next chapter. But some of the things H^+ ions are responsible for are dealt with here.

First, enzymes function best at a certain pH level. It appears the H^+ ions can influence the molecular shape of an enzyme and thus affect its ability to speed up metabolic reactions. HCl, for example, has a positive effect on the pepsin enzyme in the stomach, helping it to break down protein during digestion. Second, acids also influence the functioning of the central nervous system by acting on the cell membranes. Third, you'll recall the role of H^+ ions in energy production through ATP in the electron transport chain, discussed in Chapter 5.

Since the homeostatic control of H^+ is so important to health, let's look at how the kidney deals with it. As Figure 7.11 shows, any excess of H^+ in the blood

TABLE 7.4 pH VALUES

Solution	pH
Gastric juices	1.6
Vinegar	2.8
Coffee	5
Urine	5.5–7.5
Blood	7.35–7.45
Bile	7.8–8.6
Ammonia	11

FIGURE 7.11 Mechanisms of bicarbonate readsorption

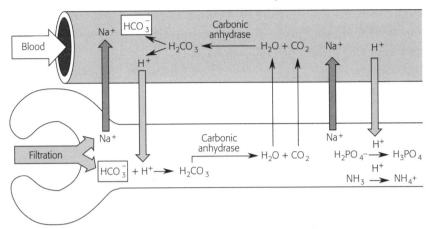

Source: *Concepts of Human Anatomy and Physiology,* 2nd Ed, Van de Graaf, K.M. and Fox, S.I., 1989, Wm. C. Brown, Iowa, p. 831.

is transported by diffusion into the tubules, where it is combined with hydrogen phosphate (HPO_4^{2-}), ammonia (NH_3) or bicarbonate (HCO_3^-).

The ammonium and hydrogen phosphate ions are excreted in the urine and the carbonic acid is recovered and breaks down to carbon dioxide and water.

This control of H^+ is referred to as **buffering**, and we can speak of the phosphate buffer system, or the bicarbonate buffer system. Since H^+ ions are produced by many metabolic reactions in cells, it is essential that various buffer systems are present to control the pH. Buffers are described in more detail in Chapter 8.

Mrs B.'s phosphate buffering system is not working. Because of the damage to her kidney, no osmosis or diffusion is taking place. The concentration of H^+ is constantly rising, despite the best efforts of her other buffer systems to cope. Therefore, the dialysis machine must remove these excess H^+ ions.

Acid/base buffering
The chemical homeostatic control of the levels of H^+ and OH^- in the body.

7.3 THE MEASUREMENT OF SOLUTION CONCENTRATION

If you look at the markings on the front of an IV bag of 5% dextrose solution, you might see the words 'Each 1000 mL provides approx. 277 mOsm, 800 kJoules'. The kJoule (kJ), remember, is a measure of energy. This section deals with the other two units mentioned, the 5% and the milliosmole, as well as other common measures.

Percentage solutions

Let's start with the 5% dextrose. A chemist would probably designate this as glucose, but the nursing profession still often refers to dextrose. Remembering that **per cent** (%) means 'per 100', we have:

Percent solutions
The ratio of solute to solvent expressed as a percentage; i.e. 5% dextrose = 5g dextrose/100mL solution.

$$5\% \text{ means } \frac{5 \text{ g dextrose}}{100 \text{ mL solution}}$$

To prepare this solution you would:

1. Weigh out exactly 5 g of dextrose.
2. Dissolve it in, say, 15 mL of water.
3. Add water to bring the volume up to 100 mL.

As a further example of this point, if you give 1 litre, or 1000 mL of 5% dextrose to a client, how many grams of dextrose does the client get?

$$1000 \text{ mL} \times \frac{5 \text{ g dextrose}}{100 \text{ mL solution}} = 50 \text{ g dextrose}$$

You may occasionally see dilute solutions expressed as milligram percent (mg%), which means the number of milligrams of solute dissolved in 100 mL of water.

Ratio solutions

Ratio solutions
Solution concentration expressed as the ratio of solute to solution; i.e. as 1:100.

You may sometimes see concentrations expressed as a **ratio** of solute to solvent. For example, the doctor orders a 1:1000 saline solution. That means 1 g of saline (salt) in 1000 mL of water. It is made up in the same way as the example above.

Molar solutions

Before considering the second unit on the IV bag, the milliosmole, we'll have to take the long way round and start with atoms. In Chapter 4 we mentioned that an atom is about 10^{-10} m (10 nm, or nanometres). That isn't very big and we can't easily work with one or two atoms, but that's seldom a problem because we normally deal with a great number of atoms at once; even a bacterium has millions of atoms. And, fortunately, we can find the number of atoms in any given piece of matter if we know the atomic weight of the elements that make it up.

The atomic weight (sometimes referred to as the relative atomic mass, for reasons we need not go into here) of an element is listed with the symbol of that element on any Periodic Table, and shows the weight of one atom of that element in relation to the other elements. Hydrogen is the smallest and therefore lightest atom and it has an atomic weight of 1. Oxygen has an atomic weight of 16, so one oxygen atom weighs 16 times as much as a hydrogen atom. Carbon, as another example, has an atomic weight of 12 and weighs 12 times as much as hydrogen. The importance of this lies in the fact that if we express the atomic weight in grams, then 1 g of hydrogen will contain the same number of atoms as 16 g of oxygen and 12 g of carbon. Of course, this is true for every element; the atomic weight in grams of any element contains the same number of particles. That number is called Avogadro's number and is about 6×10^{23} atoms. Therefore, 1 g of hydrogen contains 6×10^{23} atoms of hydrogen and so does 16 g of oxygen and 12 g of carbon; see Figure 7.12.

This system has turned out to be so useful that we now call that number of atoms a **mole** (abbreviated mol). A mole of hydrogen is 1 g of hydrogen and we know exactly how many atoms we're talking about, too. More precisely, from a chemist's point of view, the quantity of substance that contains Avogadro's number of units is called a mol.

We've now reached the point where we can talk about solutions again. This time, the amount of solute is measured in mols, rather than in grams or mL or percentages. A **molar solution** has 1 mole (or mol) of solute dissolved in a litre of water. For example, a 1 molar (1M) solution of Na^+ ions contains 1 mol Na^+ (23 g) in 1 litre of water. The solution is made up just as the others above were; the solute is dissolved in a small amount of water, then topped up to the 1 litre mark.

Mole
The SI unit of quantity. A mole is equal to the atomic weight (or gram molecular mass) of the element (or molecule) expressed in grams and contains Avogadro's number of atoms, about 6×10^{23}.

Molar solutions
Solution concentration expressed as moles of solute dissolved in a litre of solvent.

FIGURE 7.12 Examples of 1 mol quantities of elements

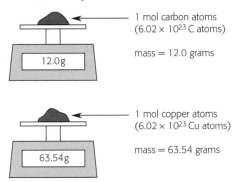

1 mol carbon atoms
(6.02×10^{23} C atoms)

mass = 12.0 grams

1 mol copper atoms
(6.02×10^{23} Cu atoms)

mass = 63.54 grams

As another example, a 2M solution of dextrose contains 2 mol of dextrose (360 g) in 1 litre of water. If you give a client 0.5 litres of this solution, the client will get:

$$0.5 \times 2 \text{ mol dextrose} = 1 \text{ mol dextrose} = 180 \text{ g dextrose}$$

SUPPLEMENT ON ATOMIC MASSES

It is not usually necessary to calculate the number of grams in a mol of a complex solute. The procedure is fairly straightforward, however, and we'll illustrate it with the dextrose example above.

Dextrose is a carbohydrate, commonly known as a sugar, with the chemical formula $C_6H_{12}O_6$. To find the molecular mass (or relative molecular mass) of this molecule, we need to add together the atomic masses of its individual atoms:

Atomic mass of carbon	=	12;	12×6	C atoms	= 72 g
Atomic mass of hydrogen	=	1;	1×12	H atoms	= 12 g
Atomic mass of oxygen	=	16;	16×6	O atoms	= 96 g
				Total	= 180 g

Not all atomic masses are whole numbers; some are decimals. For example, chlorine (Cl) has an atomic mass of 35.45. This is because elements may vary in the number of neutrons in their nuclei; see the discussion on isotopes in Chapter 12. You must take the decimal into account when adding up the atomic masses of molecules. Potassium chloride (KCl) has a molecular mass of 74.5, for example, and 1 mole weighs 74.5 g.

Milliequivalent solutions

It is becoming more and more common for nursing solutions to be expressed as molar solutions. Yet it is still possible to find some solutions, particularly of ions in the body fluids, expressed as **milliequivalents per litre, mEq/L**. (See Table 8.1 for a list of blood components.) One equivalent of an ion is found by multiplying 1 mole of the ion by the number of electrical charges on the ion. Therefore, 1 mole of sodium ions (Na^+) contains 1 Eq of sodium ions. One mole of calcium ions (Ca^{2+}) contains 2 Eq of calcium ions.

Let's see how that is used in practice. The normal level of chloride ion, Cl^-, in the blood is 100–106 mEq/L. Since the electric charge on the chloride ion is 1, then 100–106 mEq/L means the same as 100–106 mmol/L, where mmol

Milliequivalent solutions (mEq/L)
Solution concentrations that indicate the number of moles of ion solute times the number of the electric charge on the ion, divided by 1 litre of solvent.

means millimole, or 1/1000th of a mol. One mole of chloride ions is 35.4 g, so 1 mmol is 0.0354 g.

Osmole and milliosmole solutions

Recall that osmosis is the movement of solvent particles from where they are concentrated to where they are less concentrated. The concentration depends on how many solute particles there are; the more solute, the less solvent.

But think about what happens when we add some salt (NaCl) to water and it dissolves:

$NaCl \rightarrow$ $Na^+ + Cl^-$

1 particle $\rightarrow$ 1 particle + 1 particle   = 2 particles

1 mole $\rightarrow$ 1 mole + 1 mole   = 2 mole

Because solid salt, which is in the form of a lattice, breaks into its two constituent parts, there are twice as many solute particles as there were solid salt particles. Since osmosis depends on the number of solute particles, we need a unit that tells us how many particles we end up with when we dissolve a substance like salt into the solvent. The unit used is called an **osmole** and it measures the **osmolarity** of a solution. Put simply:

osmolarity = molarity × no. of particles per solute molecule

Here are two examples:

1. What is the osmolarity of a 1M dextrose solution?

   Dextrose does not break down into ions or separate atoms or molecules when it dissolves. Therefore, one molecule of dextrose gives one molecule in solution:

   osmolarity = 1M × 1 = 1 osmole

2. What is the osmolarity of a 1M NaCl solution?

   Each NaCl molecule gives two ions when dissolved, Na^+ and Cl^-:

   osmolarity = 1M × 2 = 2 osmoles

The IV bag we looked at earlier had the 5% dextrose solution as 277 mOsm. Since dextrose does not break into particles, we know that 277 mOsm = 277 mmol = 0.277 M. Since 1 M = 180 g, then 0.277 M = 0.277 × 180 g = 49.8 g. This is pretty close to the 50 g of dextrose we calculated when we looked at percentage concentrations earlier.

Osmoles and osmolarity
Osmolarity is measured in osmoles. It is the number of moles of solute times the number of particles the solute dissociates into upon dissolving. Osmoles are so named because the number of solute particles is critical for the process of osmosis.

Questions

Level 1

1. What is the osmolarity of an IV saline solution whose concentration is 0.9% NaCl?
   Answer: 300 mOsm.
2. Give specific directions for preparing 100 mL of a 5% boric acid solution.
3. Distinguish carefully between solutions, colloids and suspensions.
4. Define the SI unit of quantity, the mole. How would you make up a molar solution?
5. What does it mean when we say that the concentration of Ca^{2+} in blood is 3 mEq/L?
6. From Table 7.5, determine how many grams of Cl^- are in 100 mL of blood.
7. What is osmolarity? In what units is it measured? What is its connection with osmosis?
8. Define a milliequivalent solution.
9. What is the H^+ concentration of a solution with a pH of 5?
10. Define the two terms 'acid' and 'base'. List some of their properties.

Level 2

11. Explain in general terms how a haemodialysis machine functions.
12. What are the differences between osmosis and diffusion? Between passive diffusion and facilitated diffusion?
13. What is meant by active transport? Why does it require energy?
14. Explain the term 'osmotic pressure'. How can it prevent osmosis from occurring?
15. What is acidosis? Alkalosis?
16. What is the difference between a strong acid and a weak acid? Between a concentrated acid and a dilute acid?
17. Distinguish between buffering and neutralisation of acid/base solutions.

Level 3

18. Explain why a red blood cell will swell up and burst when placed in pure water.
19. What would happen if a 0.40M glucose solution was separated from a 0.40M NaCl solution by an semi-permeable membrane?
20. In many states in Australia, a person with a blood alcohol concentration of 0.05% is considered over the legal limit. How much total alcohol does this concentration represent, assuming a blood volume of 7 litres?

CALCULATION PROBLEM BASED ON THE CASE STUDY

After each period of dialysis, Verna B. is ordered a continuing IV infusion of 5% dextrose. The first bag containing 1 L is begun at 1400 hours at 20 drops/min. After 5 hours the rate is increased to 25 drops/min. The drip chamber delivers 20 drops/mL. What time would the next bag be hung?

Answer:

Volume A (mL) = 20 drops/min/20 drops/mL × 5 hour × 60 min/hour = 300 mL.

Volume B (mL) = 600 mL (leaving 100 mL in the bag).

Time B (min) = 600 mL/25 drops/min × 20 drops/mL = 480 min = 8 hours.

End time = 1400 + 0500 + 0800 hours = 0300 hours.

THE CORONARY CARE UNIT: THE CIRCULATORY SYSTEM

TONICITY OF SOLUTIONS, BLOOD PRESSURE AND FLOW, AND PH CONTROL

Chapter outline

8.1 **Blood as a solution.** Your objectives: to understand the concept of tonicity of solutions; and to understand the processes that affect the movement of solutes and solvents in and out of the capillaries.

8.2 **Blood and fluid pressure.** Your objectives: to understand blood pressure as an example of fluid pressure; and to understand the factors that control the movements of both static and moving fluids.

8.3 **Factors affecting flow rates of fluids in vessels.** Your objectives: to know and appreciate the health implications of blood pressure changes; to understand and perform IV delivery and flow rate calculations.

8.4 **Acidosis and alkalosis.** Your objective: to understand the role of blood gases and buffer system controls.

INTRODUCTION

Our discussion of solutions in the body is not quite complete. We still need to widen our understanding of osmosis and diffusion. The regulation of the pH levels also deserves further study, as it is a crucial part of diagnosis and treatment. The transport of solutions around the body depends on physical principles that must be understood fully. This chapter focuses on blood, using it as an example of a vital solution which is pumped through the vessels of the circulatory system. We will look at some features of those vessels, the vascular tissue of the body.

Case Study: Mr Ron J.

The coronary care unit of any modern hospital gives an excellent opportunity for observing the close relationships between science and nursing. For a start, there is obviously a lot of technology involved. Mr J. is connected by haemodynamic monitoring lines to digital readout monitors and he is being treated with mechanical ventilation. A great deal of this equipment, however, is used to supply the medical team with information. Decisions about his care depend on understanding the meaning of this information. Examine the following sections of his nursing care plan.

NURSING CARE PLAN: MR RON J.

Diagnosis: severe mitral regurgitation, substantial mitral stenosis; mitral valve replacement.

Standard care plans:
□ Haemodynamic monitoring
□ Urinary catheter
□ Artificial airway
□ Mechanical ventilation (endotracheal)
□ Chest drainage (underwater seal drain)
□ Range of motion exercises

Usual/unusual problems
Fluid balance: potential drop in cardiac output due to altered cardiac rhythm or chemical imbalance.

Nursing orders
Place on continuous ECG monitor on admission to CCU. Keep informed of serum electrolytes and blood gas levels.

Check chart for medical orders re treatment of:
☐ dysrhythmias
☐ electrolyte imbalances
☐ acid/base imbalances
☐ hypotension or hypertension

It seems that the technology of information gathering must be supported by scientific understanding. The information gathered by an ECG will be examined closely in Chapter 9. Let's start here with the serum electrolytes mentioned in the care plan.

8.1 BLOOD AS A SOLUTION

Blood is a fascinating substance for many reasons, but it is first and foremost a solution. The solvent is water and the solutes are many and varied. Table 8.1 gives some idea of the complex nature of this substance.

The words '**plasma**' and '**serum**' are virtually synonymous; both refer to the liquid part of the blood, containing the solvent and all dissolved solutes, but serum is plasma without the factors responsible for blood clotting. Plasma represents about 55% of the blood. The rest consists of blood cells and insoluble proteins.

The serum **electrolytes** are those ions such as Na^+, K^+, Cl^- and so on. They are called electrolytes because their presence allows the solution to conduct electricity. They are absolutely essential for normal cell metabolism and must be present in the serum or plasma in the right quantities. Any interruption or slowing of their delivery to the cells by the circulating blood is very serious. While Mr J. was in surgery, his heart was stopped while the mitral valve was replaced. During this time, his blood was artificially pumped around his body to maintain, among other things, electrolyte supply.

Some of the solutes in serum are not electrolytes. For example, glucose,

Plasma or serum
The liquid part of blood, containing water and all dissolved solutes. Technically, serum is plasma minus the factors responsible for clotting the blood.

Electrolytes
Ions in solution capable of conducting an electric current through that solution.

TABLE 8.1 COMPOSITION OF WHOLE BLOOD

Component	Significance
Plasma	
Water	Dissolves and transports organic and inorganic molecules, blood cells, and helps maintain body heat
Electrolytes	Normal ECF ion concentrations essential for cell metabolism
Nutrients	Energy production, growth and maintenance
Wastes	Excretion and breakdown
Proteins	Essential for osmosis; transport of lipids, ions, etc.; essential for blood clotting
Formed elements	
Red blood cells	Transport of blood gases
White blood cells	Various kinds, ranging from removal of debris to defence against foreign bodies
Platelets	Essential for clotting process

waste products such as urea and gases such as nitrogen do not dissociate into ions when dissolved. Both the electrolytes and the non-electrolytes may undergo diffusion and the water may undergo osmosis if all necessary conditions are met. Consider the red blood cell (RBC) shown in Figure 8.1(a), as it floats in the plasma.

The concentration of Na^+ ions is the same inside the RBC as outside in the plasma. The plasma is said to be **isotonic** for Na^+ with respect to the RBC. Isotonic means, roughly, 'of the same strength'. In this case there is no net

Isotonic
A solution in which the concentration of a given solute is the same on either side of the semi-permeable membrane surrounding a cell.

FIGURE 8.1(a) The ECF and the RBC are isotonic.

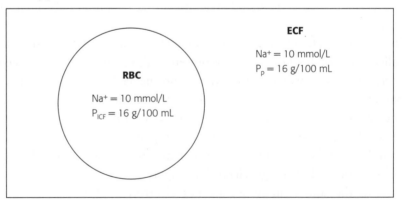

diffusion of Na^+ ions across the membrane of the RBC, assuming of course that the RBC membrane is permeable to Na^+ ions.

In Figure 8.1(b), however, the concentration of Na^+ is higher inside the RBC than in the plasma. In this case, the plasma is said to be **hypotonic** (under strength) for Na^+ with respect to the RBC. If diffusion occurs, the RBC will soon be isotonic with respect to the serum and diffusion will stop.

In Figure 8.1(c), the RBC has a lower concentration of Na^+ than the plasma has. The plasma is now said to be **hypertonic** (over strength) for Na^+ with respect to the RBC. Diffusion will move Na^+ until an isotonic state is reached.

Note how careful we have been to say that the plasma is isotonic, hypotonic, or hypertonic for Na^+ with respect to the RBC. We must always be clear what solute we are referring to, and what other solution we are comparing its concentration with. In a solution as complex as blood, any RBC may be isotonic

Hypotonic
A solution in which the concentration of a given solute is less outside the cell than inside.

Hypertonic
A solution in which the concentration of a given solute is greater outside the cell than inside.

FIGURE 8.1(b) The ECF is hypotonic with respect to the RBC.

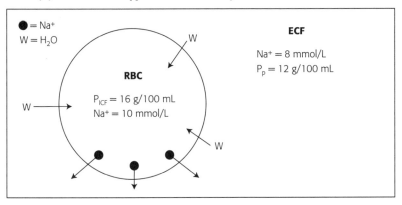

FIGURE 8.1(c) The ECF is hypertonic with respect to the RBC.

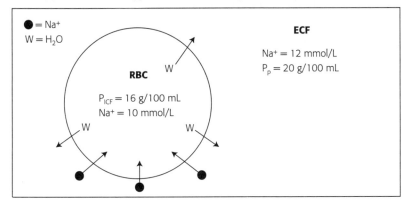

with respect to one ion, hypotonic with respect to another, and hypertonic with respect to a third.

Non-electrolyte solutes that are dissolved in water can enter through the pores, if they are small enough to pass through. But electrolytes cannot do this easily if the electric charge of the membrane repels them. It's like trying to push the north pole of a magnet through a hole surrounded by the north poles of many magnets. So electric charge may affect the rate of diffusion.

Also, since the membranes are made of a lipoprotein layer (see Chapters 4 and 6), solutes that are soluble in lipids can pass in more easily. This is true of alcohol, fatty acids and steroids, as well as dissolved gases such as oxygen and carbon dioxide.

Finally, differences in pressure can force solutes in or out of cells, influencing the rate of diffusion. This is not generally important in the body, but, as we have seen in Chapter 6, it does play a role for gases entering and leaving the lungs.

Our three diagrams above can also illustrate the possible direction of water movement, or osmosis. Recall from Chapter 7 that our definition of osmosis requires a non-diffusible solute, one that is left behind. And as in Chapter 7, let the plasma proteins play that role here.

Returning to Figure 8.1(a), let us assume that the concentration of proteins is the same in the ICF and the serum; that is, $P_{ICF} = P_p$. In this case, there will be no net movement of water (no osmosis), because there is no difference in solvent concentration (no concentration gradient).

In Figure 8.1(b), we'll assume that the concentration of plasma proteins, P_p, is less than the proteins within the cell, P_{ICF}; that is, $P_{ICF} > P_p$. In this case, the concentration of solvent is greater in the serum than in the RBC, and water will tend to enter the RBC, causing it to swell.

In Figure 8.1(c), assume that $P_{ICF} < P_p$ this time. Now, solvent concentration is higher within the RBC than in the plasma, and water will leave the RBC and move into the plasma. As a result, the RBC will become flaccid.

We have used this example to illustrate the way you need to analyse a given situation in order to determine the direction of diffusion and osmosis. You must ask yourself the following questions:

□ *Diffusion.* What solute are we interested in? Where is the concentration highest? It will move from there. Where is concentration lower? It will move towards there. Can it move through any membranes that lie between? Are there any other factors that will prevent it moving (e.g., electric charge)?

❑ *Osmosis*. Are the three conditions for osmosis present (concentration difference, membrane, non-diffusible solute)? Where is the solvent concentration highest? Water will move from there. Where is solvent concentration lower? Water will move towards there. Are there any other factors that may prevent it moving (e.g. osmotic pressure)? Don't forget that, for osmosis, you can determine where the solvent concentration is greatest by seeing where the concentration of the non-diffusible solute is smallest.

Summary

The main factors affecting passive diffusion and osmosis are:

1. concentration difference between ECF and ICF
2. pore size and size of solute molecule
3. electric charge of solute and membrane or ICF
4. solubility of solute in lipids
5. pressure differences between ECF and ICF
6. the presence of carrier molecules (see Chapter 7)
7. the concentrations of the non-diffusible solute

One of our objectives was to try to collate all we have said so far about osmosis and diffusion, to explain the movements of substances in and out of the capillaries. Consider the diagram of a capillary in Figure 8.2.

Note first that the blood in the capillary has a higher overall amount of solutes throughout its length than the surrounding interstitial fluid does. However, there is a change in the relative amounts of nutrients and wastes as we move from the artery end to the vein end. Nutrients and important electrolytes needed by body cells are in greater concentration in the capillary than they are in the interstitial fluid, so they move out to the cells by diffusion (which may be either passive diffusion or facilitated diffusion). At the same time, waste products enter the capillary because these are in a higher concentration in the interstitial fluid than they are in the capillary.

How about the movement of water? There are two things at work here, one trying to move water out and one trying to move water in. First, look at the values of the hydrostatic or water pressure. At the artery end of the capillary blood enters with a pressure of about 35 mm Hg, caused of course by blood pressure. It drops down to about 15 mm Hg by the time it reaches the vein end, due to resistance to its flow (discussed later in this chapter). These pressures are

FIGURE 8.2 Movements of H$_2$O and solutes in a capillary

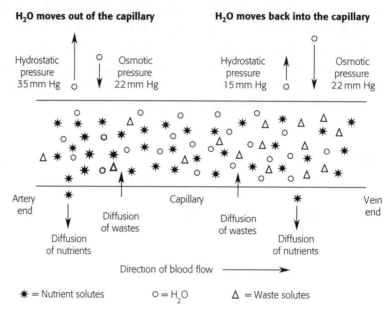

both higher than the hydrostatic pressure of fluids outside the capillary and so they tend to push water out. If this hydrostatic pressure was the only factor operating, water would tend to leave the capillaries and the blood would become more and more concentrated.

This is counteracted by osmosis. Since the blood in the capillaries is more concentrated than it is in the interstitial fluid, water tends to move by osmosis into the capillary, from higher to lower water concentration. More specifically, the relatively large colloidal proteins in the capillary are responsible for this osmosis. The value of this inward water pressure is about 22 mm Hg along the entire length of the capillary.

Putting both factors together, we can see that water tends to leave a capillary at the artery end, where the outward, hydrostatic pressure is greater than the inward water pressure (due to osmosis) by 13 mm Hg. But water then tends to enter the capillary at the vein end, where the water pressure due to osmosis is greater than hydrostatic pressure by 7 mm Hg.

Since there is a difference of about 6 mm Hg between the outward pressure and the inward pressure, water still tends to accumulate in the interstitial region, leading to oedema. This water is, however, normally returned to the circulatory system by the lymphatic system. So we can see that the movements of water and solutes into and out of the capillaries depends on factors such as:

1. The presence and concentration of proteins in the plasma; any condition that tends to change these concentrations (such as blood loss) affects osmotic flow.
2. Changes in blood pressure that influence hydrostatic pressure in the capillaries; this can result from many factors, such as changes in blood vessel diameter (vasoconstriction and vasodilation), hormones and injury.
3. Changes in the concentrations of other blood components such as nutrients and electrolytes affect the rate of diffusion of these substances; if severe, they may affect osmosis as well.

8.2 BLOOD AND FLUID PRESSURE

So far we've discussed Mr J.'s blood as a solution, with solutes and solvents and, in general terms, the delivery of these solutes to where the cells require them. It is time to be more specific about this and ask: how is this solution moved around the body? To answer that, we need to think of Mr J.'s blood as a fluid being pumped through tubes. If that sounds like hydraulic engineering, the two processes do have some things in common. For example, Mr J. is having his blood pressure continuously monitored. Any changes signal serious problems in the delivery of nutrients and in the carrying away of toxic wastes.

Let's return to the meaning of pressure. We defined it as force/area, measured in pascals (Pa). A small area can therefore lead to a high pressure, as in a hypodermic needle; the tip has such a small area that little force is needed to push it through the skin. Pressure is measured with a manometer, of which the sphygmomanometer for measuring blood pressure is an example.

Recall too the information from Chapter 3 about Pascal's Principle, which technically states that 'any change of pressure in an enclosed fluid is transmitted equally to all parts of the fluid'. Please note that it applies only to static fluids, ones that are not flowing. The blood is contained in the arteries, veins and capillaries, so it too is an enclosed fluid. It pushes against the walls of these blood vessels, creating a pressure. However, the blood is rarely if ever static, so Pascal's Principle, while useful, cannot give us a complete account of blood pressure and flow. It's the science of moving fluids we have to call on now. First consider Figure 8.3, a diagram of a hollow cylinder filled with fluid.

How much fluid flows through a tube like this? The formula is:

volume of fluid/time = pressure difference/resistance to flow

volume in mL/s = $(P_2 - P_1)/R$

FIGURE 8.3 Section of a blood vessel

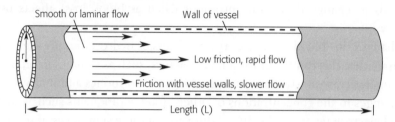

The greater the pressure difference, the greater the volume of flow. The pressure difference is responsible for the force pushing the fluid along. (In the case of the human body, the contractions of the heart are the cause of this pressure and the pressure difference is that between maximum systolic pressure and minimum diastolic pressure.) The pressure difference between one end of the tube and the other depends on two things:

1. Friction of the fluid with the walls of the tube; the greater the friction, the greater the pressure difference between the ends.
2. Turbulence of the flow; smooth or laminar flow means lower friction, whereas turbulent flow means high friction.

The smaller the resistance, the greater the volume of flow. Three things affect the resistance:

1. the radius of the tube (r)
2. the length of the tube (L)
3. the viscosity of the fluid (v)

If we put the three together in an equation we have

$$R = \frac{8 \times v \times L}{\pi \times r^4}$$

The value of these equations is that they allow us to see the relative importance of the factors involved. In this case, the most important influence on the resistance is the radius, because it is r^4. If we double the radius, the resistance decreases by $2^4 = 16$ times. However, if we halve the radius, the resistance increases by 16. This is a crucial factor when we remember that narrowing the arteries in certain heart conditions must greatly increase the resistance to the flow of blood.

Putting all of the above into one equation we have a relationship that is known as Poiseuille's Law:

$$\text{volume of fluid/time} = \frac{(P_2 - P_1) \times \pi \times r^4}{8vL}$$

Normal flow rates for a healthy adult are about 100 mL/min in a coronary artery. If that artery's radius is decreased by 20%, the new flow rate drops to 41 mL/min. To get it back to 100 mL/min, the pressure would have to rise to 244 mm Hg. Since normal systolic blood pressure is about 120–140 mm Hg, this is obviously much too high. Such pressure could cause the artery to burst.

A quick note about viscosity of the blood: because of the presence of red blood cells, it's fairly viscous, about four times as syrupy as water. The dissolved and undissolved solutes also make a difference, but their presence is far less important. In the tiny capillaries the red blood cells have to line up to pass through and the viscosity drops to half its value in the arteries. However, this is counterbalanced by the slower speed of blood through these tiny vessels, as little as 1 mm/s. This means that anything that changes the number of RBCs per mL of blood changes the viscosity—for example, dehydration, anaemia or polycythaemia, loss of blood through injury, or the addition to the blood of cell-free intravenous fluids.

Two paragraphs previously we used the term 'systolic pressure'. The contraction of the heart acts just like your fingers squeezing a water balloon. Imagine that the balloon has a hole at one end and the fluid comes spurting out—in this case, straight into an artery called the aorta. The period of contraction is when the pressure is raised the most and this pressure is called the **systolic pressure** ('systole' means to contract). When the heart relaxes between contractions, the pressure drops to a lower value called the **diastolic pressure**. The surge of higher systolic pressure is transmitted throughout the arterial system (Pascal's Principle again!) and is felt (and sometimes seen) as the pulse.

Systolic pressure
Highest blood pressure, a result of heart contraction.
Diastolic pressure
Lowest blood pressure, during the relaxation of the heart muscle between contractions.

Cardiac output

The rate at which blood is pumped around the body is known as the **cardiac output**. Strictly speaking, this is hard to measure because fluids are constantly moving in and out of the blood vessels by osmosis, so it is taken to be the amount of blood pumped by the left ventricle into the aorta each minute.

Cardiac output
The amount of blood being circulated around the body; for an adult, about 5 litres/min.

cardiac output (C.O.) = heart rate × stroke volume
C.O. (litres/min) = beats/min × litres/beat

For a healthy adult,

$$\text{C.O.} = 72 \text{ beats/min} \times 0.07 \text{ litres/beat}$$
$$= 5 \text{ litres/min}$$

This value is fairly constant, despite the changes in blood pressure that occur around the body. Look at the diagram of the pressures in the circulatory system in Figure 8.4.

Why are there these pressure differences, if Pascal's Principle is true? Well, to get any flow at all, there must be pressure difference, $P_2 - P_1$. This pressure difference partly depends, remember, on the friction of the blood with the walls of the tube. The resistance to flow also greatly depends on the radius of the tube and its length. Consider the diagrams of a river flowing, in Figures 8.5(a)–(c).

In Figure 8.5(a) the river flows with steady speed, delivering 1000 litres/min. At the point where it enters the narrow gorge, those 1000 L/min have to get through a smaller space, so the water pressure goes up, as does the speed. If that was blood going from an artery into a thin-walled capillary, the capillary would burst.

In Figure 8.5(b) the river divides into thousands of tiny channels. The total area of all these channels is much greater than the width of the river. Each

FIGURE 8.4 Pressures in the circulatory system.

Notice the general reduction in circulatory pressure in the system circuit and the elimination of the pulse pressure in the arterioles.

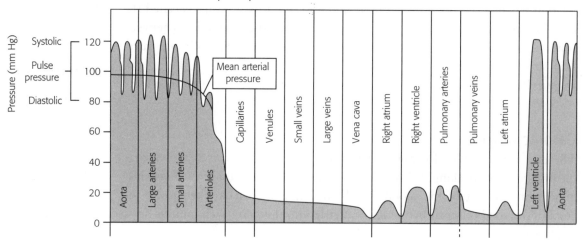

Source: *Fundamentals of Anatomy and Physiology*, 2nd Ed, Martini, F., 1992, p. 690. © Reprinted by permission of Pearson Education Inc., Upper Saddle River, N.J.

FIGURE 8.5(a) Aerial diagram of a river carrying 1000 L/min.
As it enters the narrow gorge, the speed of the water increases.

FIGURE 8.5(b) As the river divides into many small channels, the water's speed in each
decreases.

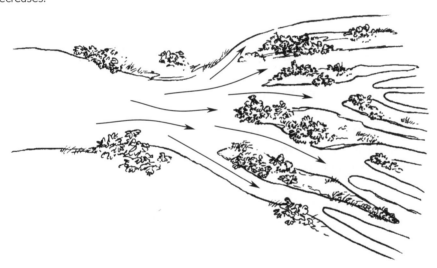

channel carries only a small fraction of the 1000 L/min, so the pressure in each
is much less. In the body, the total area of the capillaries is about 2000 cm^2, while
an artery is only about 3 cm^2, so both the pressure and speed of the blood
through them is less.

In Figure 8.5(c) the channels rejoin and flow into a wide lake. The small
pressures of the channels don't combine and since the wide lake has even less
resistance to flow, the pressure drops even further. Similarly, when the capillaries
rejoin the larger veins, the blood pressure drops, even to zero.

FIGURE 8.5(c) The separate channels rejoin into a lake, and the water's speed drops to nearly zero.

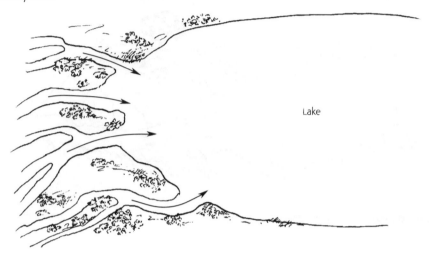

Lake

8.3 FACTORS AFFECTING FLOW RATES OF FLUIDS IN VESSELS

Mr J. was having problems with his mitral valve. Mitral stenosis refers to a decrease in the opening size of the valve, so less blood can enter the left ventricle from the left atrium during each contraction. This leads to an increase in blood pressure in the left atrium (from a normal 0 mm Hg to over 20 mm Hg in some cases). This in turn increases the pressure in the pulmonary veins that lead into the left ventricle, referred to as pulmonary hypertension. The high pressure in the pulmonary circulation interferes with the exchange of gases and with osmosis, and eventually leads to higher pressure throughout all vessels of the circulatory system. The right ventricle has to pump harder against this extra pressure and it is usually right ventricle failure that is the cause of death.

Pressure in the blood vessels

Mr J.'s mitral valve replacement will relieve the dangerously high pressures that are backing up from his left atrium. To see why high blood pressure is a problem in the arteries, we need to reconsider our long hollow tube; see Figure 8.6.

In Figure 8.6 T = thickness of artery wall

S_1 = stress along the length of the artery

S_2 = stress around the circumference of the artery

FIGURE 8.6 Stresses on the blood

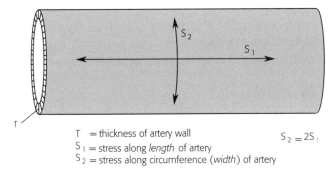

T = thickness of artery wall
S_1 = stress along *length* of artery
S_2 = stress along circumference (*width*) of artery

$S_2 = 2S_1$

The two stresses, S_1 and S_2, are caused by the blood pressure pushing on the walls of the artery. But they are not equal.

$$S_1 = \frac{rP}{2T} \quad \text{where } r = \text{radius of artery, } P = \text{blood pressure}$$

$$S_2 = \frac{rP}{T} \quad \text{therefore, } S_2 = 2S_1$$

If you recall the last time you fried a sausage, which way did it split open? Along the length, rather than around the middle, because the stress around the circumference (i.e. the width) is twice that along the length. The same thing happens to the blood vessels. They stretch both in length and in width as the blood pulses in, but the stretch in width is much greater. Even so, the increase and decrease in length of the capillaries in your brain with each heartbeat could be very worrying. So the body keeps them slightly stretched (under positive tension) to minimise any changes in length. Changes in width are our main concern.

Blood vessels are slightly elastic and can normally cope with expansion in width. However, if the radius narrows because of illness or the build-up of deposits, the stress can get too large.

For example, suppose that an artery has, initially: $r = 1$ cm, $T = 0.1$ cm and $P = 100$ mm Hg or 13.3 kPa. Therefore the stress will be:

$$S_1 = \frac{(1 \text{ cm}) \times (13.3 \text{ kPa})}{2 \times 0.1 \text{ cm}} = 66.5 \text{ kPa}$$

Now assume that, as a result of disease, $r = 0.9$ cm, $T = 0.3$ cm. If the artery holds the stress at 66.5 kPa, we can rearrange the equation to find out what the blood pressure rises to:

$$\frac{P}{r} = \frac{S_1 \times 2 \times T}{0.9 \text{ cm}} = \frac{66.5 \text{ kPa} \times 2 \times 0.3 \text{ cm}}{0.9 \text{ cm}} = 44.3 \text{ kPa} = (333 \text{ mm Hg})$$

This is clearly too high. It is quite likely that the artery will try to expand to allow the stress to drop below 66.5 kPa. If it can't, it will burst, in what is called an aneurism.

Pressure in IV delivery systems

Mr J. is on an IV drip that delivers glucose and antibiotics and, perhaps, anticoagulants. We now have some idea of the factors that affect that delivery:

1. The pressure difference, $P_2 - P_1$, is due this time to gravity. The pressure is determined by the height of the IV liquid above the point where it enters his arm. Mathematically, we say that pressure and height are proportional. The IV pressure must be enough to overcome blood pressure, but not strong enough to make the vein burst. For glucose and saline, a pressure of only 9–12 kPa is needed, as blood pressure in the veins is low. When nurses change the height of an IV bag to change the rate of flow, they are doing so by changing the pressure:

   pressure = height × density of liquid

   Since the density of the contents of the IV bag doesn't change, only the height is important.

2. The viscosity of the intravenous fluid is another factor. For glucose and saline, this is close to that of pure water, whereas blood is about three times more viscous. As the equation above shows, the pressure is also proportional to the density of the fluid and density is the ratio of mass to volume. The density of pure water, for example, is 1 g/mL. Density is related to viscosity, as dense liquids are usually more viscous.

3. The length and diameter of the tube are important. Remember our model of the river flowing into the narrow gorge; the same applies to IV delivery tubes.

Flow rate calculations

Nurses are required to be proficient at drug and drug-delivery calculations, even though many of the techniques have been made automatic. IV delivery systems in modern hospitals dispense the right amount of solution at the correct rate.

Nonetheless, there is no substitute for human alertness and understanding when monitoring clients on such treatments. In this section, we'll see how these calculations are done.

Example 1

A typical flow rate problem is as follows (all these figures are quite realistic):

What flow rate is required to administer 500 mL of dextrose 5% in water over a period of 8 hours from a set delivering 15 drops per millilitre?

The answer is simply calculated from a formula:

$$\text{rate (drops/min)} = \frac{\text{volume (mL) to be delivered} \times \text{drops per mL}}{\text{time (minutes)}}$$

For our example:

$$\text{rate} = \frac{500 \text{ mL} \times 15 \text{ drops/mL}}{8 \text{ hours} \times 60 \text{ min/hour}} = 16 \text{ drops/min (rounded off)}$$

Example 2

Ampicillin has been ordered to be infused at a rate of 100 mg/hr via a set calibrated to 60 drops/mL. The infusion contains 500 mg of ampicillin per 500 mL. What flow rate is required?

Since there are 500 mg of ampicillin in 500 mL of solution, there must be 100 mg of ampicillin in 100 mL of solution. So we need to deliver (infuse) 100 mL of solution every hour to the client:

$$\text{rate} = \frac{100 \text{ mL} \times 60 \text{ drops/mL}}{60 \text{ min}} = 100 \text{ drops/min}$$

A few more of these calculations are included in the questions at the end of the chapter.

The amount and type of IV solution delivered to Mr J. have to take into account several things. His age and size are important, because the elderly (and infants) cannot metabolise as much fluid as well as adolescents and mature adults can. His cardiovascular condition must also be kept in mind. His heart is not yet ready to pump at full strength for even short periods. Too much fluid would force the heart to work harder to pump the excess and there would also be a problem with oedema.

8.4 ACIDOSIS AND ALKALOSIS

Another item mentioned in Mr J.'s nursing care plan was blood gas levels. As we recall from Chapter 7, one of these gases, CO_2, is involved in possible acid/base imbalances, also mentioned in the care plan. Let's have a closer look at it here.

Carbon dioxide in the blood

Human beings generate about 200 mL of CO_2 per minute, because it is continuously produced by all cells in all tissues. It diffuses out into the blood and is transported to the lungs; see Figure 8.7.

Recall that there are three main ways it is carried by the blood:

1. As dissolved CO_2, just like soda water. This creates a partial pressure, p_{CO_2}, of 46 ± 4 mm Hg, which is higher than that in the lungs, 40 mm Hg. Therefore, CO_2 flows out from the blood, into the lungs and then out into the atmosphere.

2. Once dissolved in the plasma, a small amount of CO_2 combines with amino groups (NH_2) of proteins to form carbamino groups. Only about 0.5 mmol of CO_2 is carried by this system. However, once inside a RBC, much larger amounts of CO_2 can combine with haemoglobin to form other types of

FIGURE 8.7 Carbon dioxide transport in the blood

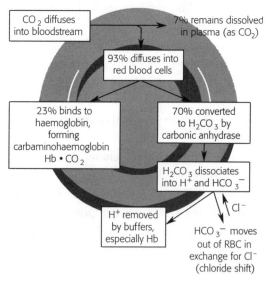

Source: *Fundamentals of Anatomy and Physiology*, 2nd Ed, Martini, F., 1992, p. 774. © Reprinted by permission of Pearson Education Inc., Upper Saddle River, N.J.

carbamino groups. The CO_2 attaches to haemoglobin that is not carrying any oxygen. In this form, it can be carried to the lungs, where the higher amounts of O_2 present knock the CO_2 off the haemoglobin and it leaves the body.

3. The most significant transport of CO_2 occurs when it is in the form of bicarbonate ion, which is formed by the following reaction in the red blood cells:

$$CO_2 \ + \ H_2O \ \underset{}{\overset{\text{carbonic anhydrase}}{\rightleftarrows}} \ \underset{\text{carbonic acid}}{H_2CO_3} \ \rightleftarrows \ H^+ \ + \ \underset{\text{bicarbonate ion}}{HCO_3^-}$$

Notice that both of these reactions are reversible; the arrows point in both directions. When there is a lot of CO_2 present, as there is when the blood picks it up from the cells, the amount of HCO_3^- inside the red blood cell increases. That is, the reaction goes to the right more than to the left. By diffusion, that bicarbonate ion then moves out into the plasma. The plasma carries it to the lungs.

At the lungs, there is a lot less CO_2 than there was deep in the body, so the reaction starts to go into reverse; that is, it goes to the left more than to the right. The bicarbonate ion adds an H^+ to make carbonic acid, which then breaks down to H_2O and CO_2. The CO_2 is then free to leave via the lungs; see Figure 8.7. In summary:

At the cells: $CO_2 + H_2O \ \rightarrow \ H_2CO_3 \ \rightarrow \ H^+ \ + \ HCO_3^-$
made by cells

At the lungs: $HCO_3^- + H^+ \ \rightarrow \ H_2CO_3 \ \rightarrow \ CO_2 \ + \ H_2O$
leaves via lungs   used by body

Acid/base balance

You spotted that H^+ in the above reactions, of course, so you're alert to the fact that we have an acid to worry about. As we mentioned in Chapter 7, it is necessary to maintain acid/base homeostasis in the body. Therefore, there must be a buffer system available to deal with that H^+ ion; and, as mentioned on Mr J.'s nursing care plan, it's something the nurse needs to understand.

Buffer systems;
see Chapter 7.

Examples:

Bicarbonate buffer system
Chemical homeostatic control of H⁺ ion concentration using the dissociation of carbonic acid, H_2CO_3.

Haemoglobin buffer system
Chemical homeostatic control of H⁺ ion concentration using haemoglobin to transport H⁺ to the lungs.

Phosphate buffer system
In cells, using the dissociation of $H_2PO_4^-$ ions.

Protein buffer system
In cells, where amino acids accept or release H⁺ ions.

In fact, so important is this homeostasis that the body has several **buffer systems** available for use. In Chapter 7 we looked briefly at one that operates via the kidney. The first one we'll consider here is known as the **bicarbonate buffer system**. Fortunately, it's fairly straightforward and is already outlined in the equation above. As discussed, carbon dioxide, when released from the cells, reacts with water to form carbonic acid. This then dissociates into hydrogen ion and bicarbonate ion. Note that it is reversible; that is:

☐ Any increase in reactants (CO_2 in this case) shifts the reaction to the right. This means more products (H_2CO_3 in this case) are made than reactants are.
☐ Any decrease in products also shifts the reaction to the right.
☐ Any increase in products shifts the reaction to the left.
☐ This means more reactants are made than products.
☐ Any decrease in reactant also shifts the reaction to the left.

Under conditions of acidosis, there is an excess of H⁺ ion (increase in product). The body contains a significant amount of bicarbonate ion, produced in the body and also taken in by the diet as sodium bicarbonate, $NaHCO_3$. This bicarbonate ion moves around the body by diffusion and can soak up excess H⁺, by shifting the equation to the left:

$$HCO_3^- + H^+ \rightarrow H_2CO_3 \text{ (carbonic acid)}$$

The carbonic acid catabolises at the lungs into CO_2, which is breathed out and H_2O, which is reused.

Under conditions of alkalosis, there is a shortage of H⁺. The carbonic acid then breaks down into the required H⁺ and HCO_3^-, by shifting the equation to the right (decrease in product):

$$H_2CO_3 \rightarrow H^+ + HCO_3^-$$

This can be taken even further. The bicarbonate ion itself can also break down, releasing yet another H⁺ ion:

$$HCO_3^- \rightarrow H^+ + CO_3^{2-}$$

This reaction is also reversible, though it's only shown going to the right in this example.

Putting all this together in one reaction, we have:

$$CO_2 + H_2O \rightleftarrows H_2CO_3 \rightleftarrows H^+ + HCO_3^- \rightleftarrows H^+ + H^+ + CO_3^{2-}$$

This bicarbonate buffer system is the most important one in the extracellular fluid. The lungs and the kidneys can control the concentrations of carbonic acid and bicarbonate in the blood, as we'll see shortly.

A second, also very important, buffer system is called the haemoglobin buffer system. This takes place in the RBC, so this system operates in their intracellular fluid. Any haemoglobin that has shed its oxygen can pick up an H^+ ion and transport it to the lungs:

$$Hb + H^+ \rightleftarrows HbH^+$$

This reaction is reversible. The HbH^+ lets go of the H^+ at the lungs, where there is lots of O_2, because O_2 replaces H^+. The released H^+ eventually reforms water and carbon dioxide.

A third control of acid/base balance is the phosphate and protein buffer systems. The phosphate system works mainly in cells:

$$H_2PO_4^- \rightleftarrows H^+ + HPO_4^{2-}$$

The protein system, which is also most effective in cells, works because amino acids (the units of proteins) can accept or release H^+ ions; see Figure 8.8.

This system is relatively slow and is most useful for dealing with the acids produced in the cell by metabolic activity. The kidneys have control over the excretion or retention of the H^+, the bicarbonate ions and the phosphate ions.

Regulation of acid/base balance by the lungs

The relevance of all this for Mr J. should now be clearer. The buffer systems are extremely important homeostatic mechanisms for acid/base control. They are tied to the amount of CO_2 in the body, which in turn is tied to the rate of breathing. For example, chemoreceptors (i.e. receptors that detect the concentrations of certain chemicals) in the brainstem are alert to any changes in the partial pressure of CO_2 in the blood. Small changes in this p_{CO_2} lead to changes in pH of the cerebrospinal fluid. A message is sent to the respiratory centres to change alveolar ventilation (Chapter 5) by a change in breathing rate.

FIGURE 8.8 Amino acid buffers.

Amino acids can either accept a hydrogen ion or donate one, depending on the pH of their environment.

Neutral pH

pH > 7
amino acid acts
like an acid

pH < 7
amino acid acts
like a base

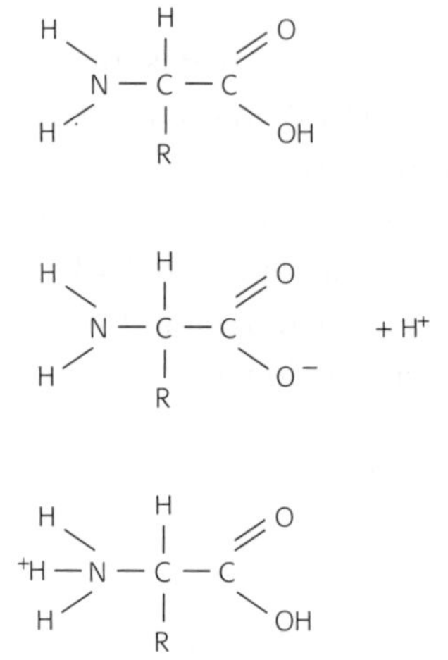

For example, suppose there is an increase in p_{CO_2} and therefore an increase in H^+ ions (can you see why?). The lungs are signalled to increase alveolar ventilation. A more rapid breathing out of CO_2 at the lungs removes carbonic acid from the blood (why?). This carbonic acid loss is a decrease in reactant and it is replaced by the combination of hydrogen ions and bicarbonate ions. This obviously removes H^+ from the blood, changing the pH (increasing or decreasing pH?). This is a classical example of a negative feedback situation, as illustrated in Figure 8.9.

This situation also shows that the CO_2 level in Mr J.'s blood must be closely monitored while he's on artificial ventilation, and afterwards as well. If there is hypoventilation, the total production of CO_2 by his body will exceed the CO_2 he breathes out. The pH of his body fluids will fall. He will try to breathe faster as a result, but this takes work by his muscles. His muscles will therefore start producing even more CO_2 as a waste product. Some of the excess H^+ will be buffered by the kidneys, but production is usually greater than that lost by buffering. This condition of sustained low pH is called **respiratory acidosis**. Its symptoms are depressed central nervous system function, somnolence, confusion and sometimes unconsciousness.

Respiratory alkalosis is far less common, caused as it is by chronic hyperventilation. Higher than normal levels of carbon dioxide are flushed from the body, leading to a loss of H^+ ion (can you see why?) and a raising of pH. The

Respiratory acidosis
Excess H^+ ion concentration due to hypoventilation, causing a build-up of CO_2.

Respiratory alkalosis
Low levels of H^+ ion, caused by hyperventilation flushing CO_2 from the body.

FIGURE 8.9 Flow diagram of a homeostatic response to increased p_{CO2}

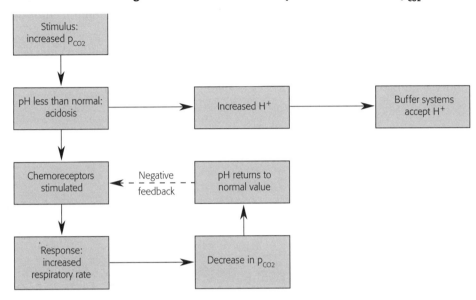

usual symptoms are agitation, dizziness, light-headedness and shortness of breath. In more severe cases, tetany and coma may result. The main cause of respiratory alkalosis in hospitals is through iatrogenic (which means 'resulting from the actions of physicians') hyperventilation. Clients who, like Mr J., are intubated, may suffer sharp drops in p_{CO2}, which can lead to increases in pH of the cerebrospinal fluid.

Non-respiratory or **metabolic acidosis** is slightly misnamed since it doesn't have much to do with metabolism. There are two cases: (1) bicarbonate ion is lost to the body by things such as diarrhoea, certain renal diseases, or some drug treatments; (2) there is an excess production of acids in the body (perhaps due to metabolic causes), or they are taken in through drug therapy. This may be due to uraemia, for example, or diabetes. In either case, the pH falls and the body tries to capture the excess H^+ with bicarbonate ion as usual. If this is not completely successful, respiration increases, though complete return to homeostasis is rare. Acute cases lead to cardiac dysrhythmia and central nervous system dysfunction. Chronic, low-level cases usually adjust to a pH level lower than normal, with no serious ill-effects.

Metabolic alkalosis is due either to an increase in alkali, or loss of acid. The latter is more common, caused by chronic vomiting (loss of HCl) or the use of diuretics, which flush metabolic acids from the body and in this case there are

Metabolic acidosis
Loss of bicarbonate ion from the body, or an excess of acids taken into, or produced by, the body.

Metabolic alkalosis
An increase in alkali into the body, or loss of acid by the body (e.g. by vomiting).

excess bicarbonate ions in the fluids, which soak up virtually all the remaining H^+ ions and raise the pH. It is often treated by prescribing ammonium chloride (NH_4Cl), which breaks down to HCl and NH_3 in the liver.

The chemical homeostasis of acids and bases is critical for health. When it is up to the medical staff to monitor changes in blood gases and pH levels, a clear understanding of the chemical processes taking place is necessary.

Questions

Level 1

1. What are serum electrolytes?
2. How does the amount of oxygen carried by the haemoglobin change as the partial pressure of oxygen increases? As it decreases? What significance does this have for the cells and hypoxia?
3. How is carbon dioxide carried from the cells to the lungs?
4. Where does blood flow most quickly in the circulatory system? Most slowly? Where is blood pressure highest? Lowest?
5. Define the terms 'systolic' and 'diastolic' as they apply to the heartbeat.

Level 2

6. Name two causes for metabolic acidosis. How does the body try to restore homeostasis?
7. Why must an isotonic solution be given during a blood transfusion?
8. What would happen to a red blood cell if it found itself in a hypotonic solution? A hypertonic solution?
9. Describe the reaction between an acid and a base and the bicarbonate buffer in the blood.
10. What are the effects on the blood pH of hypoventilation? Why does it occur?
11. What factors affect the rate of blood flow through an artery or vein? Why is the size of the radius so crucial?
12. What factors affect the delivery rate of IV solutions?
13. Explain the physical dangers associated with mitral stenosis. How are they normally treated?

Level 3

14. Discuss why an increase in blood flow rate does not necessarily mean an increase in blood pressure at the same time.
15. Why are the blood vessels normally kept under tension?
16. What types of buffers are present in the blood? How do they work?
17. If you increased your heart's stroke volume by 10%, but kept your cardiac output at 5 litres/min, what heart rate would you have?
18. What flow rate is required to administer 600 mL of saline in 8 hours, from a set delivering 60 drops per millilitre?
    Answer: 75 drops/min.

19. A client suffering anxiety begins to hyperventilate. What effect does this have on the P_{CO2} of the blood? Explain the changes in pH that may result and the mechanisms by which the body attempts to restore homeostasis.

20. A blood transfusion is infusing at 25 drops per minute and the set is calibrated at 15 drops per mL. What is the hourly infusion rate, in mL?

Answer: 100 mL/hour.

21. The nursing care plan calls for an infusion of 1.5 L of normal saline, with an antineoplastic drug, over 6 hours. For a regular drip, drop factor of 10, calculate the rate of flow. *Note*: A drop factor of 10 means that it requires 10 drops from the drip chamber of the IV set to deliver 1 mL.

Answer: 42 drops/min.

THE INTENSIVE CARE UNIT: ELECTRICITY AND MAGNETISM IN THE BODY

VOLTAGE, CURRENT, RESISTANCE, CONDUCTORS AND INSULATORS

Chapter outline

9.1 **Electricity and magnetism.** Your objective: to understand these from an atomic level perspective.

9.2 **Voltage, current and resistance.** Your objective: to define and use these terms appropriately.

9.3 **Conductors and insulators in the body.** Your objectives: to define these terms correctly; to apply them to matters of electrical safety; and to understand the meaning and implications of both macro- and microshock.

9.4 **The diagnostic and therapeutic uses of electricity and magnetism.** Your objective: to understand and appreciate the applications of electricity and magnetism for monitoring, diagnosis and treatment.

INTRODUCTION

Chapter 8 mentioned the need to monitor clients. This means to gain constant information about their internal and external condition. In some cases this can be done quite simply with the basic skills of measuring and recording such things as temperature, pulse and respiration. However, technology now allows the nurse to measure far more complex, and formerly inaccessible, functions of the body. Many of these technological advances depend on both electricity and magnetism. As well, the body itself uses electricity to carry out many essential tasks, such as beating the heart, contracting the muscles, sending signals around the body and thinking. In this chapter we turn our attention to the basic science behind the use of electricity and magnetism for diagnosis, treatment and normal body function.

Case Study: Mr Richard S.

Richard S. is recovering from a 2° myocardial infarction. This indicates difficulty with the flow of electrical energy through the conducting fibres of the heart. At the age of 63, a smoker and overweight, he seems a high-risk candidate for such trouble. Fortunately, this time the loss of blood supply to the heart was localised and brief. Surgery was not required, but pacemaker wires have been implanted to ensure smooth, regular heartbeat. He is being monitored in the intensive care ward until the medical team is assured that his treatment has removed him from a life-threatening position. The nursing goals for Mr. S. are shown to you by the CNC of intensive care. They include: to restore myocardial tissue perfusion, reduce chest pain, provide adequate nutrition, and provide information about the illness or treatment. In this case, providing such information means knowing something about the principles of electricity as they apply to excitable tissue in the human body.

Clinical note: Nurses looking after Mr S. would be expected to obtain and report haemodynamic parameters reflecting decreased fluid volume: BP less than 90/60 mm Hg; CO less than 4 L/min; Central venous pressure (CVP) less than 2 mm Hg; and mean arterial pressure (MAP) less than 70 mm Hg.

9.1 ELECTRICITY AND MAGNETISM

To understand what electricity and magnetism are, it is necessary to return to atoms. Recall the picture of the atom given in Chapter 4. There we discussed a central nucleus, consisting of positively charged protons and uncharged neutrons, circled by a number of negatively charged electrons. It is the electrical attraction between the positive and negative charges that holds the electrons in their orbits.

This attraction is an example of electrostatic force, as defined in Chapter 4. Recall that it is a force between charged particles that may be either attractive or repulsive. The strength of the electrostatic force in any given situation depends on how many positive or negative charges are present and on how close together they are. The important thing is, the charges will be pushed and pulled around by the electrostatic force. If they aren't attached to something, they will move.

It is also true that the negative charge on an electron is exactly as strong as the positive charge on a proton. So when they get together, they cancel each other out:

proton and electron = neutral; +1 + −1 = 0

If there are more negative particles (anions) than positive particles (cations) at some location, then the total electric charge there is negative. For example:
15 anions and 10 cations = charge of −5

$$-15 \quad + \quad +10 \quad = \quad -5$$

Of course, this applies to any anions or cations.

Now, let's apply these facts to the following situations. Let's start with something simple. In this case, what would you expect to happen?

$H^+ + OH^-$

The hydrogen ion and the hydroxide ion will attract one another, move together and form a neutral molecule of water, H_2O. Next, something a little larger in scale:

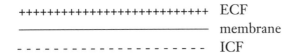

++++++++++++++++++++++++++++ ECF
———————————————————————— membrane
- ICF

The positive and negative particles are separated by a membrane, like that surrounding a cell. We say that the ECF has a net positive charge with respect to the ICF. Alternatively, we can say that the ICF has a net negative charge with respect to the ECF. If the particles are free to move, they will move towards each other, through the cell membrane (if it's permeable to those ions). This movement is the result of both diffusion and the electrostatic attraction between the oppositely charged ions. At the finish, there will be equal numbers of positive and negative ions on each side of the membrane:

$$+ - + - + - + - \quad \text{ECF}$$
$$- + - + - + - + \quad \text{ICF}$$

Now we can say that there is no net electric charge around this section of the membrane. It is electrically neutral.

But is it magnetic? **Magnetism** is caused by electrons spinning around their axes, as the earth does around its poles. One pole of the electron is called the north magnetic pole and the other is called the south magnetic pole. Normally, the north pole of one electron cancels out the south pole of any nearby electron. Overall, therefore, the body is magnetically neutral. Except under unusual circumstances or when using certain kinds of monitoring equipment, magnetism can be ignored.

Magnetism
Caused by the movement of electric charges, it results in the establishment of north and south magnetic poles. Living organisms are generally magnetically neutral, as the two poles always occur together.

9.2 VOLTAGE, CURRENT AND RESISTANCE

Richard S. is connected to an external pacemaker, a technique usually referred to as cardiac pacing. The medical team hopes to control the rate at which the heart beats, allow time for minor ischaemic damage to heal and get the heart's natural beating pattern back on rhythm.

In the healthy heart, what we call the heartbeat is of course a contraction of the heart muscle. This contraction requires an electrical stimulation. The cells that provide that stimulation are located in the sinoatrial node (usually called the SA node). Let's have a closer look at one of the cells there.

Figure 9.1 presents a sketch of a typical cardiac muscle cell in the SA node.

On the ECF side of the membrane there is an excess of Na^+ and Ca^{2+} ions. This condition is described, remember, as one of net positive charge on that side of the membrane. Now, while there is also an excess of K^+ on the ICF side of the membrane, there are also lots of negatively charged ions, such as certain proteins, hydrogen phosphate and bicarbonate. As a result, the inside of the cell

FIGURE 9.1 Cardiac muscle cell, with a resting membrane potential of about –90 mV

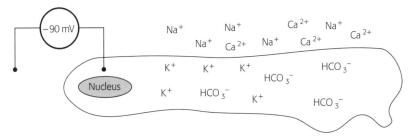

has a net negative charge. At this stage the cell is just like a miniature electric battery, like the ones that run a torch or radio. These too have a positive end and a negative end.

If you think about it, it's pretty unlikely that all those positive Na^+ and Ca^{2+} ions just happen to be in the one place like that. Normally they repel one another, as well as being attracted by the negative ions. This situation is so unexpected that it is reasonable to suppose that the cell has used some of its energy, in some form of active transport, to deliberately separate those positive and negative ions. The situation is similar to pulling apart two magnets that have been joined north pole to south pole. You have to use at least a little energy to separate them. If you put them close together again, and let them go, they would fly back together. In scientific terms, the magnets first store up the energy you used to pull them apart (as potential energy; see Chapter 3) and then they use it up (as kinetic energy; Chapter 3) by rushing back together.

The same argument applies to a SA node cell. The cell uses energy from its supply of ATP to separate like charges from each other. The cell has a sodium–potassium pump, which actively pumps Na^+ ions out and K^+ ions in. By opening or closing its membrane channels, the cell can also change its permeability to different ions, thus changing the rate of diffusion. Using both its pump and its control over permeability, the cell can carefully regulate the type and amount of charge on either side of its membrane.

In Figure 9.1 the cell is shown connected to an instrument for measuring the amount of energy represented by the separation of the charges. The separation has resulted in what is called negative electric potential—negative because, by convention, we always refer to the net electrical charge inside the cell; electric because we're dealing with the separation of electrically charged ions; and potential because there is stored (or potential) energy in the ions, given to them by the cell's ATP.

Volts (V)

Technically,
energy/charge. The
energy comes from the
work required to separate
oppositely charged
particles (e.g. the work
done by neurones using
active transport) and the
charge is the charge on
the ions that are moved.
This is also referred to as
electric potential, as the
energy is in the form of
potential energy (e.g. the
neuron's negative electric
potential of −70 mV).

How much energy are we talking about here? Electrical energy is usually measured in units called **volts**. In the SI system that we are using, a volt is defined as energy/charge. This means we need to ask: how much energy is given to each unit of charge, positive or negative? In the case of a cell in the SA node, the energy is about −90 millivolts (mV); (it's −90 because it's a negative electric potential, remember). When the SA cell 'fires' it stops holding the ions apart and allows them to rush either in or out. That stored potential energy is now available as electrical energy. As a result of the ions moving across the membrane, a new situation occurs in the space of milli-seconds; see Figure 9.2.

First, there is momentary neutrality, then the inside of the cell becomes positively charged. There is now a positive electric potential of about +20 mV. This voltage is high enough to stimulate surrounding heart muscle to contract and so the heart beats. The SA node cell then gets to work, pumping the ions back to their starting-points, restoring the −90 mV potential, ready to fire again. As you know, it does this about 50 to 80 times a minute, every minute of your life, once for every heartbeat. Cells capable of 'firing' like this are called autorhythmic. Cardiac muscle cells are one example, and neurones are another. They are special types of what is called 'excitable tissue', tissue capable of responding to an electrical stimulus. The details of how autorhythmic cells function are considered more fully when we look at the function of a neurone later in this chapter.

Return to that SA node cell (Figure 9.1) just before it fires. The ions are sitting quietly on either side of the membrane. This is an example of **static electricity**. The separated charges are not moving. Static electricity is very common. For example, walking across carpets can rub electrons off the atoms of the carpet fibres and they accumulate on you and your clothing. You become

Static electricity

An accumulation of
excess positive or
negative charges, which
can discharge through a
spark; it is particularly
dangerous in microshock.

FIGURE 9.2 When a cardiac muscle cell in the SA node fires, Na⁺ and Ca²⁺ ions enter the cell through channels, reversing the membrane potential from −90 mV to +20 mV.

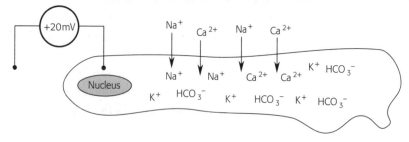

negatively charged and release those electrons through a little spark when you touch the door handle.

When the ions in the SA node cell rush together in response to the attraction of the electrostatic force, this is referred to as current electricity; a small electric **current** is flowing. An electric current is simply a collection of charged particles (ions or electrons) moving from one place to another. If we could count how many of these charged particles pass us by in a second, we could measure the strength or size of the current. Current, therefore, is charges/second. In the SI system, charges/second is an ampere, usually shortened to amps (A). The symbol for electric current is I. In the heart, currents of less than 0.05 mA are common. Domestic appliances use 240 volts in Australia (compare this with the heart's voltage of about –90 to 20 mV) and 5–8 amps.

Of course, the ions never move completely freely from place to place. There is always some resistance. In the case of a SA node cell, this resistance comes from the cell membrane, the atoms and molecules in the solutions on either side and the disturbing influence of other charged particles in the immediate environment. The effect of all these is to interfere with the easy flow of electric current. This interference is called electrical **resistance** and is measured in units called ohms. The ions lose some of their energy in fighting through this resistance. We take advantage of this fact in a light globe. The resistance in the filament is so high that the electrons lose their 240 V of energy, turning it into heat and light. The symbol for electric resistance is Ω, the Greek letter omega.

The amount of current, I, that flows depends on how much energy the particles have (voltage, V) and how much resistance, Ω, there is. Mathematically, this is expressed as a ratio:

current = voltage/resistance

In symbols: $I = V/R$

For Richard S.'s condition we can now understand what is meant by a pacemaker current of 5 mA, a skin resistance of 10 000 Ω and a pacemaker voltage of 10 V. If any of the pacemaker values change, or if his skin resistance goes down, he may experience dysrhythmia and/or cardiac arrest. This is because the muscle cells in his heart contract when they receive the electric current spreading out from the pacemaker's electrode. They are triggered off by it and in turn trigger those cells next to them. In this way, a wave of contraction sweeps out from the SA node over the atria. When the electric current reaches the AV node, it sends the message along through the ventricles. Too much or

Current (I)
The number of charged particles that move past a given point in a second, measured in amperes. In relation to voltage (which supplies the energy for the movement) and resistance (which tends to prevent the movement), $I = V/R$.

Resistance (R)
Any opposition to the free movement of electric charges, measured in ohms (Ω).

too little current, voltage or resistance will therefore seriously affect heart contractions.

Nerve cells and electricity—a more detailed look at the movement of charged particles in cells

So far we have presented one of Mr S.'s heart cells as an example of electricity in the body, but the same principles apply to any cell capable of being autorhythmic. Since the nerve cells (neurones) are especially important for electrical activity in the body, let's look at them in some detail. As we do so, we'll keep referring to ideas developed earlier, such as diffusion, the electrostatic force and the structure of a cell membrane.

Figure 9.3 shows a section through a neurone at rest. It is, you recall, in a state of negative electric potential, which in a nerve cell is also called its **resting potential**.

At this time there are clear differences in the concentrations of the ions between the inside (ICF) and the outside (ECF) of the cell. As for the SA node cells discussed above, the movements of these ions are controlled by three factors:

Resting potential
The state of voltage of a neuron when it's at rest, just before being stimulated, usually about −70 mV.

1. *Diffusion*; they tend to go from high concentrations to low.
2. *The electrostatic force*; they tend to move towards particles of opposite charge and away from particles of similar charge.
3. *The permeability of the membrane*; some ions pass through the membrane more easily than others.

FIGURE 9.3 Resting membrane potential of a neurone, with ion concentrations in mmol/L

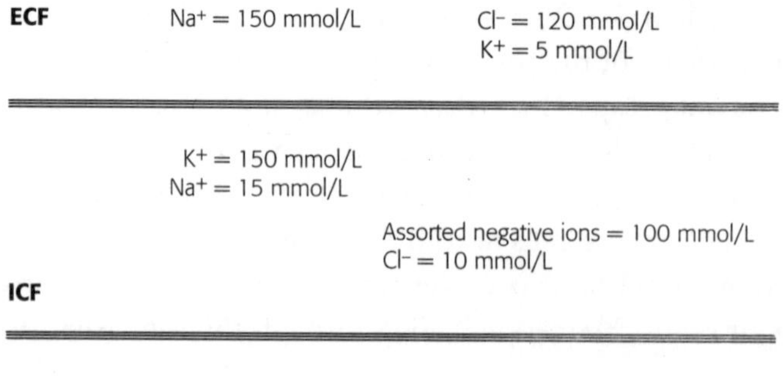

ECF $Na^+ = 150$ mmol/L $Cl^- = 120$ mmol/L
                                                                    $K^+ = 5$ mmol/L

$K^+ = 150$ mmol/L
$Na^+ = 15$ mmol/L

Assorted negative ions $= 100$ mmol/L
$Cl^- = 10$ mmol/L

ICF

These factors work together to create the resting potential of the nerve cell, as follows.

The cell pumps Na^+ ions out and K^+ in, until the ECF concentration of Na^+ equals the ICF concentration of K^+. This is an example of active transport (discussed in Chapter 7) because the ions are being moved against their concentration gradient. The membrane, however, is more permeable to K^+ than to Na^+. This means K^+ tends to diffuse out into the ECF more quickly, until there is enough total positive charge in the ECF to repel them. All these positive ions in the ECF make the outside of the cell more positive than the inside, where there are still lots of Cl^- ions and other negative, non-diffusible ions such as certain proteins.

The Cl^- ions in the ICF are pulled towards the outside by the electrostatic force, and also diffuse through. The membrane is quite permeable to Cl^- ions. But since the cell continually pumps K^+ back in, they can help to hold in the Cl^- by lessening the outward electrostatic pull. You can see that this pumping is necessary to maintain a resting potential of −70 mV, which would quickly become electrically neutral if diffusion was allowed to go unopposed. In this state, the cell is said to be polarised. One way to think of this term is to remind yourself of a simple dry cell, with a positive end (positive pole) and a negative end (negative pole). A fully charged battery can be also thought of as polarised in this sense.

A neurone can respond to a stimulus by changing the electric potential from −70 mV to +30 mV in about 1 millisecond. This change is called **depolarisation** and the sudden variation in the membrane voltage is called an **action potential**. From resting to action, there is a great change in the concentrations of ions in the ECF and ICF, as shown in Figure 9.4.

The stimulus, which may be a touch, light or sound, for example, causes a sudden, large change in the permeability of the membrane to Na^+ ions. Why? Because the stimulus makes large pores (called voltage-sensitive sodium channels) open in the membrane and Na^+ ions pour into the cell, pushed by diffusion and by their attraction to the negative ions in the ICF. These pores are designed so that only Na^+ ions can pass through, controlled by little gates made of protein molecules. This movement of Na^+ ions represents a small electric current. And as these positive charges enter the ICF, they cause the voltage to change. The pores soon snap shut, because they are influenced by the surrounding voltage (which is why they are called voltage-sensitive).

With all these extra Na^+ ions in the ICF, there is now a positive electric potential of +30 mV across the membrane. This potential closes the Na^+

Depolarisation
The term for the response of a neuron to an action potential, in which it changes its electric potential from −70 mV to +30 mV in about 1 ms.

Action potential
The stimulus that makes a neuron fire or polarise.

FIGURE 9.4 The generation of an action potential

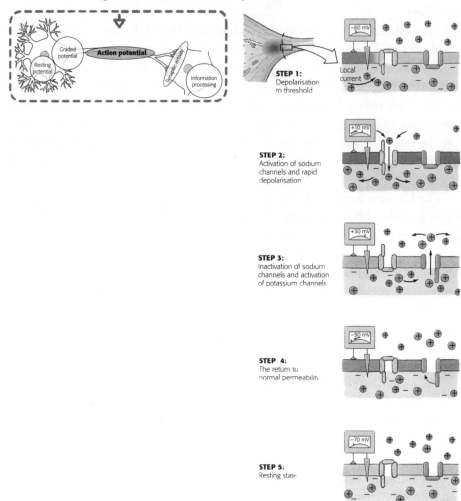

Source: *Fundamentals of Anatomy and Physiology*, 5th Ed, Martini et al, 2001, p. 381. © Reprinted by permission of Pearson Education Inc., Upper Saddle River, N.J.

channels, and causes some voltage sensitive K⁺ channels to open. Again, diffusion and electrostatic forces combine to push the K⁺ ions out of the cell. So many of them leave, in fact, that the electric potential now switches over to about –75 mV. This changeover is called repolarisation. After many repetitions of the change from polarisation to **repolarisation**, the cell then uses active transport, in the form of a potassium–sodium pump, to move the Na⁺ back outside and pull the K⁺ in, restoring the resting potential. Those voltage-sensitive channels open and close so fast that only a tiny fraction of the available

Repolarisation
The process of returning the neuron to its resting potential after it has depolarised

ions move in and out when the cell fires. A neurone would have to fire many thousands of times before all the Na⁺ and K⁺ ions could change places. The whole event is so fast that a neurone can generate an action potential as often as 1000 times a second. These events are shown by Table 9.1.

These movements of ions, which first occur at the point where the stimulus is, then act as a stimulus to nearby sections of the cell membrane. As a result,

TABLE 9.1 SUMMARY: ACTION POTENTIALS

STEP 1: Depolarisation to threshold

☐ A graded depolarisation brings an area of excitable membrane to threshold (−60 mV).

STEP 2: Activation of sodium channels and rapid depolarisation

☐ The voltage-regulated sodium channels open (sodium channel activation).
☐ Sodium ions, driven by charge attraction and the concentration gradient, flood into the cell.
☐ The transmembrane potential goes from −60 mV, the threshold level, toward +30 mV.

STEP 3: Inactivation of sodium channels and activation of potassium channels

☐ The voltage-regulated sodium channels close (sodium channel inactivation). This occurs at +30 mV.
☐ Voltage-regulated potassium channels are now open, and potassium ions move out of the cell.
☐ Repolarisation begins.

STEP 4: Return to normal permeability

☐ Voltage-regulated sodium channels regain their normal properties in 0.4–1.0 msec. The membrane is now capable of generating another action potential if a larger-than-normal stimulus is provided.
☐ The voltage-regulated potassium channels begin closing at −70 mV. Because they do not all close at the same time, potassium loss continues and a temporary hyperpolarisation to approximately −90 mV occurs.
☐ At the end of the relative refractory period, all voltage-regulated channels have closed and the membrane is back to its resting state.

Source: *Fundamentals of Anatomy and Physiology*, 5th Ed, Martini et al, 2001, p. 382. © Reprinted by permission of Pearson Education Inc., Upper Saddle River, N.J

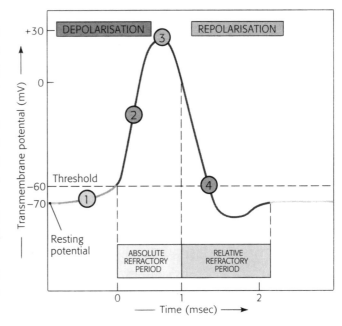

these sections also start to depolarise and ions begin to rush in and out of their voltage-sensitive channels. The next section of membrane is then stimulated and on it goes. This results in a pulse or wave of firings (called a nerve impulse) as the action potential moves along the neurone.

To summarise these steps in sequence:

1. A neurone, before it fires, is at a resting potential of –70 mV. We now assume it is stimulated.
2. The voltage-regulated sodium channels open and Na⁺ ions rush into the ICF, due to both diffusion and electrostatic attraction. The membrane potential changes over to +30 mV.
3. The sodium gates close and almost simultaneously the potassium gates open. The K⁺ ions rush out of the cell, again due to diffusion and electrostatic forces. This loss of positive charges shifts the membrane potential back towards the resting potential.

The potassium channels close rather slowly and for a brief period the membrane potential is about –75 mV, which is lower than the resting potential. This condition is called hyperpolarisation. It comes to an end when the potassium gates finally close, active transport has done its job and the resting potential of –70 mV is restored.

9.3 CONDUCTORS AND INSULATORS IN THE BODY

Earlier in this chapter we mentioned the movement of ions as an electric current and said that there is always some resistance to that movement. To complete our discussion of the terms for electricity in the human body, we need to say more about this.

Any substance through which ions can move relatively freely is called a **conductor**. The wires of Richard S.'s pacemaker provide a good example of a conductor that is a solid. Obviously, they have to be conductors, to get the electricity from the power supply to the heart. The interstitial fluid surrounding the cells of his heart is full of electrolytes (remember, these are ions capable of conducting an electric current). This fluid, then, is an example of a conductor that is a liquid.

By contrast, any material through which ions cannot travel at all, or only with great difficulty, is called an **insulator**. The plastic covering of the pacemaker

Conductor
A material (solid, liquid or gas) that allows the free movement of charged particles (current electricity) through it. For example, *electrodes*, which can be used to detect the presence of charges, or conduct it to any particular point.

Insulator
The opposite of a conductor; it prevents or slows down the flow of current electricity.

itself is an insulator, as is the covering around the wires that plug the monitor into the wall. Our skin is a good insulator, too. When it's dry it can have a resistance of over 100 000 Ω, but this can drop to less than 1000 Ω when the skin is wet.

As you would expect, neurones are good conductors of electricity. A nerve impulse can move down a neurone at up to 130 m/s. The neurone's membrane itself is also a good conductor, except when it's covered with myelin, an insulating lipid, which acts like the plastic coating on the monitor wires.

The movement of electrons or ions requires conductors. In the language of science, these conductors form an electric circuit, or pathway. Any breaks or gaps in this circuit act like a resistance and stop the electric current from moving. This is one reason why nurses are urged not to run trolleys over the cables in the intensive care ward; there is a small chance that this will interrupt the flow of current.

In the human body, the electrolytes act like conductors in a circuit. If an electric current can get through the resistance of the skin, it can travel fairly freely throughout the entire body. This is exactly what happens when we receive an electric shock, which if severe enough can cause death. Three things determine how dangerous this is:

1. the amount of current (number of electrons or ions) passing through the body
2. how long the shock lasts
3. which parts of the body are affected

Burns can result from electric shock. The tissues of the body tend to heat up, just like the filament in a light globe, because they are better resistors than the interstitial fluid that surrounds them. The main problem, though, is that the electric current stimulates muscles to contract and neurones to fire. In particular, muscles controlling breathing and heartbeat can be affected. A current of merely 30–200 mA passing through the body can make the heartbeat become uneven (dysrhythmia).

If Richard S. was not on an external pacemaker, he would be in no real danger of electrocution. His skin resistance, like ours, is high enough to prevent normal amounts of electric current from entering his body. However, the gel used with his monitoring electrodes is designed to lower the resistance of the skin at that point. Of course, if he somehow came in contact with an exposed 240 V electric wire, then electrocution is possible. That sort of situation is called **macroshock**, because high voltages and strong current are involved.

Macroshock
Electrocution involving high voltages and currents, passing through the skin.

However, the pacemaker wires provide an electrical pathway straight to his heart. If by some accident a current as small as only 60 mA passed down those wires, his heart could suffer ventricular fibrillation. That is, the muscles contracting his ventricles would begin to spasm uncontrollably. This is an example of **microshock**, or low voltages and very small current. They can occur if a member of the health team carrying static electricity simply touched the pacemaker wires, for example, or if the client touched two different pieces of electrical equipment (see Table 9.2).

The major protection of the client against macroshock is probably the responsibility of technical staff of the hospital, yet it would pay for the nurse to realise the dangers to the body of static electricity build-up. The way to prevent any possibility of microshock is to make sure that you and all electrical equipment are **earthed**, or **grounded**. That is, make sure there is a circuit carrying any excess charge out of the body or equipment and into the earth. Touching the doorknob after walking across the carpet is one way of earthing, though not the most pleasant. Prevention is the better technique, of course, and explains why hospital floors are rarely carpeted.

Microshock
A type of electrocution involving very small voltages and very small currents, which gain entry to the body by avoiding the skin's protective high resistance

Earthing, or grounding
Providing a pathway to the earth for charged particles; providing a safety control of static and current electricity

TABLE 9.2 EFFECTS OF ELECTRIC CURRENT ON BODY FUNCTION

Current (mA[1])	Effect	Voltage required[2] (V)
1	Just detectable	10
5	Maximum harmless current	50
10–20	Sustained muscular contraction ('can't let go')	100–200
50	Pain, exhaustion	500
100–300	Ventricular fibrillation; fatal if continued	1000–3000
6000	Sustained ventricular fibrillation; temporary respiratory paralysis; burns	60 000

[1] = 1 second duration
[2] = skin resistance of 10 000 ohms

9.4 THE DIAGNOSTIC AND THERAPEUTIC USES OF ELECTRICITY AND MAGNETISM

In the case of Richard S., it's clear that electricity is being used not only to help him (via his pacemaker), but also to monitor his progress. While the nurse may not need to know the technical details of the monitoring equipment used in a modern hospital, there must be some understanding of the principles behind their use. Otherwise, the results displayed by the monitor, or the therapy provided by the pacemaker, cannot be promptly and intelligently interpreted. The next sections look briefly at some common applications of electricity and magnetism in health-care settings.

The electrocardiogram (ECG)

In discussing the **ECG**, we can put together a lot of information about the heart and electricity.

The cardiac cycle

When the SA node cells fire, the electric current spreads throughout the muscle cells of both atria and they begin to contract. Blood is forced into the ventricles. The atria then begin to relax. About 0.16 seconds after the SA node has fired, the electric current reaches the AV node. The firing of the AV node begins the contraction of the ventricles, forcing blood out of the heart. This strong contraction is the systolic phase and blood pressure readings are at their highest. Now the ventricles start to relax. The period of time in between contractions is referred to as the diastolic phase, when blood pressure readings are lowest.

The heart cells are contracting because they are depolarising in response to the electric current reaching them from the SA and AV nodes. Since it is happening in many thousands of cells, this depolarisation creates an electric current strong enough to be detected through the skin. The detection is done by electrodes, usually metal, which are sensitive detectors of the presence of charged particles. As each depolarisation and repolarisation is detected, it is displayed on the oscilloscope screen of the ECG. In Figure 9.5 each wave of electric current, whether depolarisation or repolarisation, shows up as a spike on the screen. Each spike has been given a letter to identify it.

The ECG electrodes are not placed just anywhere on Richard S. They are carefully placed to produce the clearest signal to the monitor and to reveal the greatest amount of information about his heart. Though the number and position of the electrodes varies, depending on what the health team is trying to find out, it is usual to place them in a triangle, known as the Einthoven Triangle, around the body, as shown by Figure 9.6.

If the electric current that started off in the SA node is travelling towards an

Electrocardiograph (ECG)
A machine that detects the electrical activity of the heart and displays it on an oscilloscope screen and as a graph on paper. The display is known as an electrocardiogram, and can be recorded on paper for later analysis.

FIGURE 9.5 An electrocardiogram.

An ECG printout is a strip of graph paper containing a record of the electrical events monitored by electrodes attached to the body surface. The placement of electrodes affects the size and shape of the waves recorded. This is an example of a normal ECG using three electrodes in a standard position (left and right wrists, and left lower leg). The enlarged section indicates the major components of the ECG and the measurements most often taken during clinical analysis.

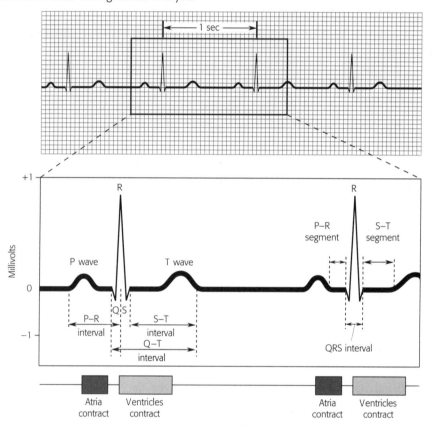

Source: *Fundamentals of Anatomy and Physiology*, 2nd Ed, Martini, F., 1992, p. 657. © Reprinted by permission of Pearson Education Inc., Upper Saddle River, N.J.

electrode, the monitor shows this as a spike that points upward, like the P or T wave. If the electric current is travelling away from an electrode, the monitor shows this as a spike that points downward, like the S wave. Thus we know both the amount of voltage (height of spike) and the direction of current (up or down spike). Since the amount of current depends on how many heart muscle cells are firing, the height of the spike also tells us something about the size and/or health

FIGURE 9.6 ECG electrode positions.

This most common arrangement is known as the Triangle of Einthoven.

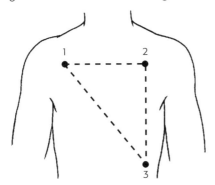

of the heart muscles. All this is summarised in the diagram of a healthy heart's ECG in Figure 9.7.

We do not see a spike resulting from the repolarisation of the SA node and the atria, because that takes place while the much greater depolarisation of the ventricles is occurring. The large QRS wave complex drowns out any other spikes.

Richard S., however, does not have a healthy heart. He has suffered a small myocardial infarction and since some of his heart cells have suffered ischaemia (deficiency of blood supply) they are short of oxygen. Under such conditions, they may start to depolarise spontaneously without waiting for the SA or AV node signal. This random firing and contraction of heart muscle is referred to as **fibrillation**. Figure 9.8 shows an example of an altered rhythm (in this case a missed beat).

Fibrillation
The random depolarisation of heart muscle tissue, due to loss of control by the sinoatrial node. It is sometimes treated with an electric current from a defibrillator.

FIGURE 9.7 A typical ECG pattern of a healthy heart

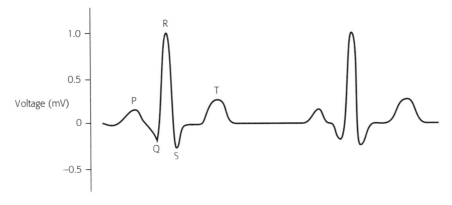

FIGURE 9.8 An ECG pattern showing an abnormal rhythm—a dropped beat.
The atria are contracting at a faster rate than are the ventricles.

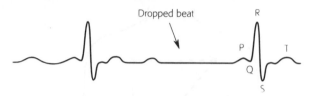

If Richard S.'s myocardial infarction had been more severe, or more prolonged, heart muscle cells might have died. Cell death is called necrosis. In that case, things are a little more complicated. There are fewer cells to depolarise, so the ECG waves are usually lower. The dead cells also interfere with the movement of the electric current across the heart. Live cells may not receive the message to fire at all, or they may receive it at the wrong time and start fibrillating. Of course, if enough cells die, the heart ceases to work at all. ECG patterns can reveal a great deal to the trained eye about the electrical and physiological condition of the heart.

The pacemaker

Pacemakers are designed to do the job of the SA and/or AV nodes. Indeed, the SA node cells are often referred to as pacemaker cells. Whether implanted in the client's body or connected to an external power supply, the pacemaker supplies regular electric current to the heart muscle (myocardium). The current passes to the heart through an electrode, usually placed in the right ventricle as shown in Figure 9.9.

Most often, there is damage to the AV node in myocardial infarction, so it is necessary to stimulate the ventricles with the pacemaker. In Richard S.'s nursing care plan, the suggested settings for the dials on the external pacemaker are as follows:

AV interval:	2 milliseconds (how long the current lasts)
Mode:	demand (pacemaker responds to SA node rate)
Current commence:	6 mA (pacemaker current)
Rate:	variable; commence at 80/min
Voltage:	adjust to threshold; 0.7 mV

The power supply for an external pacemaker, of course, is the 240 V mains supply to the hospital. For an implanted, internal pacemaker, there are now

FIGURE 9.9 The placement of a pacemaker in the right ventricle.
The power supply is often implanted in either the abdomen or shoulder.

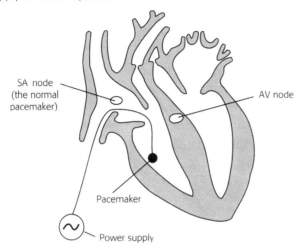

several options for power. There are long-life (2–4 year) mercury-filled batteries; piezoelectric crystals that generate electricity when pushed or pulled by the movements of the body; bio-galvanic batteries that use the electrolytes of the body; and radio-operated devices that trigger the pacemaker with radio waves sent from outside the body, just like a radio-controlled aircraft.

The defibrillator

Small amounts of electric current (0.005–100 mA) can cause ventricular fibrillation, as mentioned above. Unless this is corrected, the client can die. One treatment is to try to get the heart's natural SA node rhythm back in action. By briefly applying a large current, the ventricles can be made to contract for a short time, which stops fibrillation. When the current is shut off, the heart may then resume its normal pacing.

The machine used to supply this external current to the heart is known as a defibrillator. Its electrodes are in the shape of large, flat surfaces, often called paddles. The large area is important, for if the electric energy was concentrated in a small area, the client could be burned. The paddles use a large voltage to stimulate a large current in the heart's autorhythmic cells.

The electrical energy the defibrillator delivers to the patient is often shown on the machine in joules. For example, charging the paddles to an energy of 100 J, and delivering it in a burst lasting only 2.5 ms, represents about 1400 V.

The correct voltage to use depends on the client's condition at the time. The external current can be delivered more than once if required and the voltage adjusted when necessary.

The electroencephalogram (EEG)

Since the brain is (on one level!) a large lump of nerve tissue, it too produces detectable electric currents. Electrodes placed on the skull can detect and relay these signals to a monitoring device, the **electroencephalogram (EEG)**. Unlike the electrical pattern of the heart revealed by the ECG, the brain's activity shows no regular pattern. There are differences between sleeping and waking, but the spikes on the EEG show that electrical activity varies depending on how active the brain is. EEG information can be summarised with a diagram; see Figure 9.10.

The voltages range considerably too, from 0 to 300 µV. These are much lower than ECG voltages. Since the brain is encased in a thick layer of bone, the strength of signals is reduced.

Diagnostically, the EEG can help to find tumours in the brain. The tumour site effectively blocks some of the recognisable brainwave patterns and this disruption to the expected EEG pattern reveals the tumour's presence and location. Also, some tumours cause surrounding brain tissue to fire in quick, sporadic bursts—another clue to potential trouble. And finally, we are beginning to understand something of the condition known as epilepsy by studying the

Electroencephalogram (EEG)
A machine for detecting and displaying the electrical activity of the brain. Typical voltages recorded range from 0 to 300 µV. The display is known as an electroencephalograph.

FIGURE 9.10 Different types of electroencephalograph waves

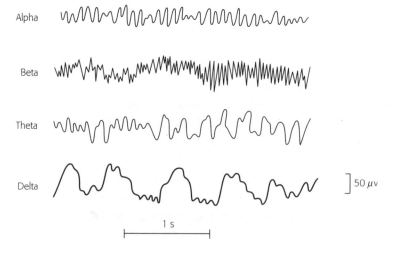

electrical activity of the brain during grand mal seizures. Epilepsy appears to be a large-scale, uncontrolled firing of large sections of brain tissue all at once. Why this occurs and how it can be prevented may eventually be understood with help from the EEG.

Electroconvulsive therapy (ECT)

The EEG is designed as a receiver of the brain's electrical signals; **electro-convulsive therapy** is designed to send electrical signals to the brain, in the hope of curing or reducing a client's condition. It is generally prescribed only for severe or chronic depression.

ECT has an interesting and sometimes controversial history. Even today, there is no firm theory about how it works. The electrical energy supplied to the brain seems to increase the production of certain chemicals by the neurones. Some researchers believe that an imbalance in these chemicals can lead to depression. The applied voltage ranges from 70 V to 110 V and lasts for only 0.1 second per pulse.

Electroconvulsive therapy (ECT)
The application of low levels of electric current to the brain in an attempt to treat depression.

Use in orthopaedics

There are many applications of electricity in nursing, but one of the more interesting ones occurs in orthopaedics. Occasionally, broken bones fail to join together and this is known as delayed union or non-union. To speed up the process, electrodes are placed either in or surrounding the bones and the current switched on. Usually, 10–18 V are used. Because of the high resistance of bone, there is only a measured 1 mV per centimetre of bone, a very safe level. For 10–12 hours a day, the bones are saturated with the electrostatic force. This enhances the production and deposition of the minerals from which bone is made, but it is not yet clear why it works.

Magnetic resonance imaging (MRI)

Also known as nuclear magnetic resonance (NMR), this technique uses the magnetic field in the nucleus of the atom. As the electrons do, protons spin; this makes them act like little magnets. The magnetic field of the detector forces these 'magnets' to line up in specific directions. The proton 'magnets' can be made to jump from one direction to another when radio waves of the right frequency strike them. These jumps reveal details of the physiological characteristics of the tissue containing the protons. Areas of the body that contain abnormalities such as tumours give a different pattern of jumps. Once they are detected by the NMR scanner, their picture is then analysed by computer.

A complete description of all the diagnostic and therapeutic uses of electricity and magnetism would require a book to itself. Let's end with two other applications—they can be looked up in the nursing literature by those who would like more information.

Electrocautery

A resistor at the end of a long probe is heated to a high temperature by an electric current. It is then used to burn small areas of tissue to stop bleeding.

Thermistors

Thermistors are thermometers. Their operating principle is rather interesting. The resistance of a wire to the movement of electric current through it depends partly on the temperature of the wire. Heat up the wire and the resistance increases. Thermistors are probe wires that can be inserted into the body. The temperature change causes a change in the resistance and this change is shown on a dial marked in degrees Celsius.

Questions

Level 1

1. Define the terms 'voltage', 'current' and 'resistance'. How are they related?
2. What does a resting potential of −55 mV mean? Where has the energy to separate the ions come from?
3. What is static electricity? How can it cause a problem in a hospital setting?
4. What voltages and currents are associated with microshock? With macroshock?
5. What is fibrillation? Why is it dangerous?
6. What are voltage-sensitive sodium channels? What do they do?
7. What is the function of a pacemaker?

Level 2

8. Describe the electrical process by which the heart muscle goes through the cardiac cycle.
9. Describe in some detail the movement of ions that occurs when a neurone transmits an action potential. Use diagrams if necessary.
10. Explain the terms 'depolarisation', 'repolarisation' and 'action potential'.
11. After a neurone has 'fired', why does it need a waiting period before it can fire again?
12. If the skin resistance is 5000 Ω and a shock comes from a power point of 240 V, what current then flows through the skin?
   Answer: 0.048 amps.
13. Why is an unearthed electrical monitor a potential danger?
14. Give an example of the use of the EEG for diagnosis.

Level 3

15. Why is it not easy to specify how many volts represent a health danger?
16. When a person suffers macroshock from a power line, they are often unable to let go of the wire. Explain.
17. What precautions should be taken to protect clients from microshock?
18. Why are clients not shocked by the voltages used in electrocautery?
19. Describe the information that is contained in each of the waves (PQRS) displayed by an ECG graph.

CALCULATION PROBLEM BASED ON THE CASE STUDY

Mr S. is ordered digoxin 0.125 mg. Available are 'Lanoxin' 250 µg and 'Lanoxin PG' 62.5 µg. It is always safer to use whole tablets than to use half or quarter tablets. Since two strengths are available, which would you give Mr S.?

Answer:

Lanoxin:

Number of tablets = 0.125 mg/250 µg = 1/2 tablet.

Lanoxin PG:

Number of tablets = 0.125 mg/62.5 µg = 2 tablets (given).

THE CLINICS PART 1: LIGHT AND SOUND IN NURSING

PROPERTIES OF LIGHT AND SOUND, VISION AND HEARING, NOISE POLLUTION

Chapter outline

10.1 **Light and waves.** Your objective: to understand the characteristics of waves; frequency, wavelength and amplitude.

10.2 **Refraction.** Your objective: to apply this property of light so that you understand how the eye focuses light onto the retina.

10.3 **Eyesight correction.** Your objective: to apply the physical principle of refraction accurately to cases of visual impairment, such as myopia.

10.4 **Electromagnetic waves.** Your objectives: to appreciate the nature and range of the electromagnetic spectrum; and to see how the different types of waves can be used for diagnosis and treatment.

10.5 **The ear and sound waves.** Your objectives: to understand the structure of the ear and its function as a receiver of sound waves; to know and apply the physical properties of sound waves relevant to human hearing; and to understand and apply the principle of resonance to the reception of sound.

10.6 **The medical applications of sound.** Your objective: to appreciate the uses of sound waves for diagnosis and treatment.

10.7 **Noise pollution.** Your objective: to understand some of the problems and physical principles associated with noise pollution.

INTRODUCTION

As human beings, we gain a great deal of information about the world outside us through our senses. This chapter looks at the characteristics of light and sound, the physical principles behind their reception by the eye and ear and some of the problems that arise in sight and hearing impairment. It also looks at some of the uses of light and sound for both diagnosis and treatment. Hopefully, the chapter will help you understand the problems faced by the sensory impaired and why we treat their conditions as we do.

Case Study: Mrs Janice D.

Mrs D. has been vision-impaired since birth. She suffers from congenital closed-angle glaucoma. The intraocular pressure of her eye is higher than normal (normal is 12–21 mm Hg) and she required surgery several years ago. She now visits the hospital's eye clinic once a month to have her intraocular pressure checked. As well, she treats her eye with 2 drops of timolol maleate twice a day (gtt 1–2, bd). These drops help decrease fluid production in the eye. The goals of her nursing management and intervention are mainly educational ones, so that the nature of glaucoma and her treatments are understood and accepted by Mrs D.

As an example of this, consider the fact that Mrs D. must have her intraocular pressure measured on a regular basis. The usual instrument for doing this is called a tonometer. In her case, the clinic uses a contact tonometer that is placed gently on the eyeball. Her eye must be given a local anaesthetic beforehand. The pressure reading is taken directly from a scale measured in mm Hg. This procedure, and her preparation for it, must be clearly understood by both Mrs D. and the nurse attending.

Clinical note: The eye drops she is using, timolol, is an example of a beta blocker. It can have significant side effects, especially in the elderly. The nurse must ensure that these side effects are looked for and monitored if they occur.
At the age of 67, Mrs D. is also beginning to show many of the age-related characteristics of hearing loss. While these do not need medical treatment, the nurse has a role to play here too. Mrs D. would benefit from having the nature of her hearing loss explained to her and should be advised about strategies that can minimise their effect on her personal and social life.

10.1 LIGHT AND WAVES

Light is not a simple phenomenon. In their study of its behaviour, physicists have found that sometimes it acts like a series of water waves and other times like a hailstorm of little particles called photons, though even these are not like atoms or particles of matter. In fact its behaviour resembles that of wave and particle, but to understand its role in health only, we'll think of light as if it behaved in the same way as water waves.

Imagine spending some time watching the waves roll in and crash on the beach. Some things are clear at once.

1. The waves are a certain size, or height above the surface of the ocean. This height is called their **amplitude**. It represents how much energy went into forming the wave out at sea. The wave, in fact, is transferring the energy of its source (such as a storm) from far out at sea to the shore. Waves carry energy. The symbol for amplitude is A, and, if we increase the amplitude of a light wave, we make the light brighter.

2. The waves come in one after another, separated by a certain distance. This separation is called their **wavelength** and is measured from the crest of one wave to the crest of the one behind it. Its symbol is the Greek letter lambda, or λ, and for light the wavelength distances are very small, measured in nanometres (10^{-9} m). Red light, for example, has a wavelength of about 7.5×10^{-7}m, which is 750 nm. These two characteristics are illustrated with the diagram of a typical wave in Figure 10.1.

3. We can count the number of waves that crash on the shore in, say, a minute. Suppose there are 10 waves/minute. This is called their **frequency**, and is symbolised as f. For light, the frequencies are measured in waves per second.

Amplitude
For a water wave, its height above level surface; for a sound wave, the size of an area of compression or rarefaction. In the case of light, amplitude is related to brightness; for sound, it is related to loudness.

Wavelength
The distance between one part of a wave to the corresponding part on the succeeding wave. For light, an indicator of colour; for sound, an indicator of pitch.

Frequency
The number of waves per second, measured in hertz (Hz). The product of frequency and wavelength is the speed of the wave.

FIGURE 10.1 Characteristics of a single-frequency, continuous wave

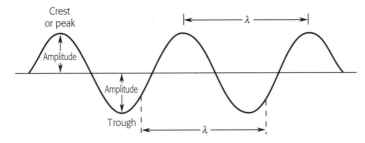

Source: *Physics*, 3rd Ed, Giancoli, D.C., 1991, p. 288. © Reprinted by permission of Pearson Education Inc., Upper Saddle River, N.J.

The unit for this is called the hertz (Hz). Light comes in many frequencies, just like water waves. Red light, again, has a frequency of about 8×10^{14} Hz.

4. The water waves come in at a varying speeds—some fast, some slow. Unlike water waves, though, light always travels at the same speed in a vacuum. There, the speed of light is about 3×10^8 m/sec. Light waves, however, do slow down when they hit something transparent, like water or glass.

If you stop and think about those water waves, there is a connection between the frequency, wavelength and speed. The faster they come into the shore, the greater the frequency. And the closer they are together (short wavelength), the greater the frequency. Mathematically, these relationships can be shown as:

speed = frequency × wavelength

or:

$$v = \quad f \quad \times \quad \lambda$$

Note that amplitude (*A*), the energy of the wave, does not appear here. Changing the energy does not affect its frequency, wavelength or speed, only the light's brightness. For example, dimming the lights in a theatre does not change their colour.

With water waves, it's the water that's moving. With light waves, however, it's a combination of electric and magnetic fields. The origin of these waves is in the movements of electrons in atoms, from one energy level (discussed in Chapter 4) to another. This movement generates alternating electric and magnetic waves which travel freely through empty space. Therefore light is known as an electromagnetic wave. What's the main difference between light waves that have different frequencies? Well, the brain interprets the different frequencies as different colours. Table 10.1 shows the frequencies and wavelengths matched to the colours as perceived by the human brain.

TABLE 10.1 RELATIONS BETWEEN COLOUR, FREQUENCY AND WAVELENGTH

Frequency ($\times 10^{14}$ Hz)	Wavelength (nm)	Colour
3.94 to 4.60	760 to 647	Red
4.60 to 5.12	647 to 585	Orange
5.12 to 5.22	585 to 575	Yellow
5.22 to 6.11	575 to 491	Green
6.11 to 7.07	491 to 424	Blue
7.07 to 7.89	424 to 380	Violet

Of course, these are not the only possible frequencies and wavelengths for electromagnetic waves; they are only the ones we can detect with the eye. Wavelengths greater than 760 nm, for example, are responsible for infrared waves. Those shorter than 380 nm give rise to ultraviolet waves. More will be said about these later. For now, keep in mind that visible light is only a small part of the total spectrum of electromagnetic waves, as shown in Figure 10.2.

It may come as a surprise to learn that radio and television waves, and X-rays too, are just electromagnetic waves of different frequencies and wavelengths.

FIGURE 10.2 The electromagnetic spectrum

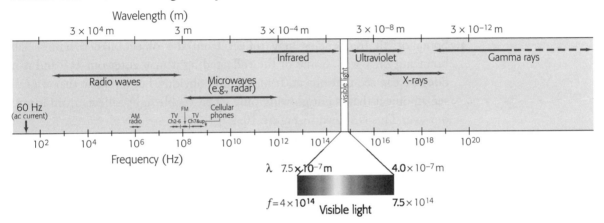

Source: *Physics*, 5th Ed, Giancoli, D.C., 1998, p. 670. © Reprinted by permission of Pearson Education Inc., Upper Saddle River, N.J.

Summary

1. Amplitude is evidence of energy and represents brightness of light.
2. Frequency is waves per second (Hz) and determines colour.
3. Wavelength is the distance between waves and also determines colour.
4. Speed of light is a constant in a vacuum.
5. Visible light is only a small part of the possible frequencies of electromagnetic waves.

10.2 REFRACTION

Visible light is usually given off by an object that is hot enough to glow. Of course, if it is glowing we can see it, so it must be visible light! Before it gets as hot as that, it still gives off infrared or heat waves, as mentioned above and in Chapter 2. If the object could be heated to a high enough temperature, it would give off ultraviolet waves too. All these waves spread out from the glowing source in all directions, as Figure 10.3 shows.

These waves travel out from the source at the speed of light, getting weaker and weaker as they go (the term for this is 'attenuated'). What happens if a light wave runs into something as it travels from the source? There are three possibilities; let's assume the light is bright enough for easy, clear seeing, and is composed of a number of different frequencies, from red to violet.

1. Some or all frequencies may be absorbed by the object it hits. If they all are, the object appears black and it generally warms up a little because of the extra light energy it absorbed. It appears black because no light from that object is reaching the eye and black is due to the absence of light.

2. Some or all frequencies may reflect from the object, like a tennis ball bouncing off a wall. Some of the reflected light may enter our eye and if it does we can see that object. It may sound obvious, but it is true that we can see an object that is not glowing only when visible light reflects from it into our eye. This is shown in Figure 10.4.

3. Some or all frequencies may pass right through the object, if the object is transparent. The light is slowed down a little and may change direction, but comes out the other end of the object and continues to travel.

Note that, whenever light falls on an object, some of the frequencies of that light may be absorbed by the object, some reflected and some pass right

FIGURE 10.3 Light waves are emitted from a glowing source in all directions.
The colour of the light is determined by the frequency, which in turn depends on the energy.

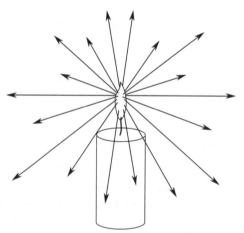

FIGURE 10.4 Light rays emerge from each single point of an object.
A small bundle of rays leaving one point is shown entering a person's eye.

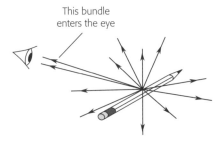

This bundle
enters the eye

Source: *Physics*, 3rd Ed, Giancoli, D.C., 1991, p. 590. © Reprinted by permission of Pearson Education Inc., Upper Saddle River, N.J.

through. This combination of possibilities accounts for the visual appearance of objects seen. Objects which reflect most frequencies look lighter; those which allow only red light to pass through are red filters, and so on.

So in order for us to see an object, some light must reflect from it into the eye, travel through the transparent material of the eye and then be absorbed by the detectors of the retina, at the back of the eye. These detectors are special light-sensitive cells called rods and cones, according to their shapes. Rods detect low levels of light, while cones are responsible for colour vision and brighter light levels. When light energy strikes either type, it causes chemical substances in the detectors to alter three-dimensional shape. Rhodopsin is the chemical in the rods and iodopsin is in the cones. This chemical change results in the stimulation of nerve fibres leaving the eye and an electrical impulse is sent to the brain, where the signal is interpreted. The anatomy of all this is shown in Figure 10.5.

Our own experiences of looking through transparent materials show that what we see through them is sometimes very distorted. If the light waves from the object we're looking at get all jumbled up as they pass through the transparent material, the image becomes blurred. As Figure 10.6 illustrates, light waves from the top of the object end up on the left, those from the left end up on the right and so on.

This implies that the light waves passing through the structures of the eye must be kept in order and this means two things: (1) the surfaces of the cornea and the lens must be smoothly curved; (2) the material in the eye must be very clear and very uniform. If the cornea and lens are not smooth and symmetrical, the condition is known as **astigmatism**. The blurred image that results can often be corrected with eye glasses or contact lenses.

Astigmatism
A condition causing blurred vision, caused by unevenness of the cornea and/or lens. Refraction does not occur uniformly.

FIGURE 10.5 Image formation by the eye.

(a) and (b) show how light from each portion of an object is focused on a different part of the retina. The resulting image arrives, (c) and (d), upside down and backwards.

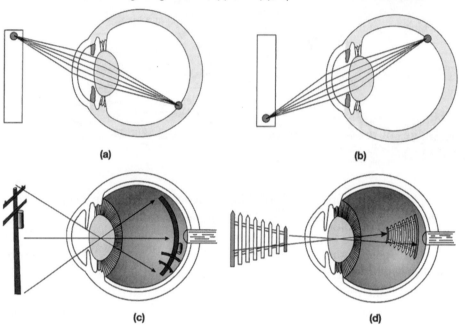

Source: *Fundamentals of Anatomy and Physiology*, 5th Ed, Martini et al, 2001, p. 549. © Reprinted by permission of Pearson Education Inc., Upper Saddle River, N.J.

FIGURE 10.6 When light travels through a transparent material with uneven surfaces, the image is distorted:

(a) clear image due to smooth surfaces; (b) distorted image due to uneven surfaces.

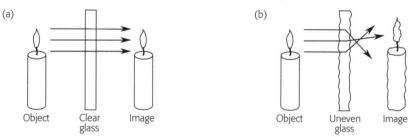

But the material in the eye must be more than clear and uniform. Light waves slow down as they go from the air into the eye, because the transparent solids and liquids of the eye are denser than air. It would be like trying to drive your car underwater; even if the engine could keep going, the denser water would greatly reduce its speed. As well as, and because of, slowing down, however, the light wave also changes direction slightly. It bends as it enters the eye; see Figure 10.7.

As the two tyres on the right side of the car enter the mud, they lose their grip on the road and that side of the car slows down. The tyres on the left side, still on solid ground, push that side of the car at the original speed. Since the left side of the car is travelling faster than the right side, the car skids around to the right. It is now facing and moving in a new direction. The same thing happens to a light wave if it enters the eye or glass at any angle other than 90°, as shown by Figure 10.8.

FIGURE 10.7 Refraction.
(a) The right tyres enter the mud, and lose grip, slowing down the right side of the car, while the left side remains at its original speed. (b) As a result, the car skids to the right, facing a new direction.

(a)
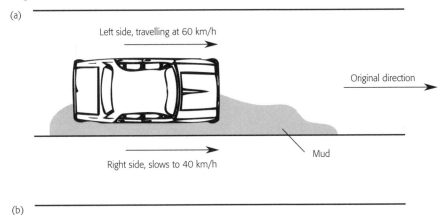

Left side, travelling at 60 km/h

Original direction

Mud

Right side, slows to 40 km/h

(b)
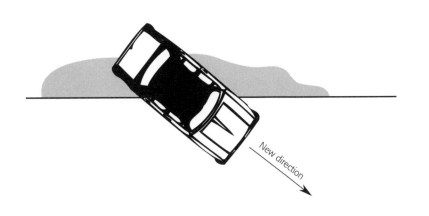

New direction

FIGURE 10.8 Refraction of light.

As the light wave enters the glass at an angle, the leading edge slows down first, pulling the whole wave into a new direction.

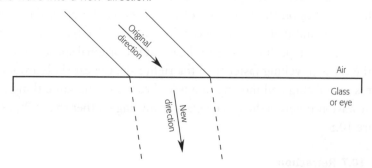

Refraction

The change in direction of a wave due to a change in speed. For light, this allows an image to be focused on the retina.

This bending of light as a result of its slowing down is called **refraction**. You can see this if you look at a straw in a glass of clear liquid—it appears bent. This is very important for understanding how the light waves reflected from an object are focused on the retina at the back of the eye; refer back to Figure 10.5.

The curvature of the eye and of the lens must be exactly correct. If it is not, the light waves are not focused on the retina. Instead, they focus either in front of it or behind it. Consider our case study again. Mrs D.'s eyes are not of the normal shape or size. Because of the increase in fluid in the eye, they are both slightly larger than usual, and the curvature of the cornea is less than normal. As a result, focusing is not as easy for her.

But for all of us, including Mrs D., focusing is made even more tricky because the eye is constantly looking at things that are near, as when we're reading, or far away, as when we glance up from the book and gaze out the window. The difference this makes is amplified in Figure 10.9.

To allow for this fact, the shape of the lens must change every time you shift your gaze from a nearby object to one further away. The lens of the eye is made

FIGURE 10.9 Accommodation

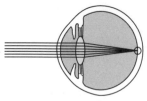

Ciliary muscle relaxed, lens flattens for distant vision

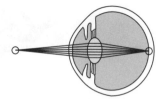

Ciliary muscle contracted, lens rounded for close vision

Source: *Fundamentals of Anatomy and Physiology*, 2nd Ed, Martini, F., 1992, p. 554. © Reprinted by permission of Pearson Education Inc., Upper Saddle River, N.J.

of a flexible material which can be stretched to make the lens flatter, or relaxed, to make the lens rounder. This is done by the tiny suspensory ligaments and the ciliary muscles. The slight change in shape (called **accommodation**) keeps the light waves focused precisely on the retina and the image is clear.

10.3 EYESIGHT CORRECTION

Unfortunately, as a result of age, or illness, or heredity, some people lose the ability to change the shape of their lens. In some cases the ciliary muscles are weakened; in some the lens itself loses its flexibility. Of course, as with Mrs D., the same condition can be caused by changes in the size of the eye itself. The distance between the lens and the retina has increased to the extent that the lens cannot accommodate enough to keep nearby objects in focus. Regardless of the cause, the result is known as either **myopia** or **presbyopia**, depending on how the lens is affected. Let's look at these two common problems separately.

Myopia

Here, the light waves from distant objects come to a focus in front of the retina. This happens either because the eyeball is too deep, or the lens cannot be made thin enough. The image is then blurred. Nearby objects, however, are seen quite clearly, so this is also called nearsightedness or shortsightedness. It can be corrected by glasses that have concave lenses, because these spread the light waves out a little before they reach the eye. Before such glasses were common, people tried to correct the problem by squinting; myopia comes from Latin words that mean 'to shut the eye'.

Hyperopia and presbyopia

In these cases, the light waves from nearby objects come to a focus behind the retina. Either the eyeball is too shallow (hyperopia) or the lens loses elasticity (presbyopia). The ciliary muscles must contract, bulging out the lens, even to focus on distant objects. They cannot contract enough, however, to allow normal vision of nearby objects. Since distant objects are clear, this is called farsightedness. It is corrected by glasses with convex lenses, since they do some of the work of the lens for it, bringing the light waves slightly together. Since the loss of lens elasticity often happens normally with age, it's no surprise than presbyopia comes from words meaning 'old man's eye'. Both these conditions are illustrated by Figure 10.10.

Accommodation
The changing in shape of the lens of the eye to allow it to focus the light from objects at varying distances from the eye.

Myopia
Near or shortsightedness, caused by either an eyeball that is too deep or the inability of the lens to elongate fully (accommodate); images are focused in front of the retina. It is corrected with concave lenses.

Presbyopia and hyperopia
Farsightedness, caused either by an eyeball that is too narrow (hyperopia) or the loss of elasticity by the lens (presbyopia); images are focused behind the retina. They are corrected with convex lenses.

FIGURE 10.10 Nearsighted eye and farsighted eye

(a) A nearsighted eye, which cannot focus clearly on distant objects, can be corrected by use of a diverging lens. (b) A farsighted eye, which cannot focus clearly on nearby objects, can be corrected by use of a converging lens.

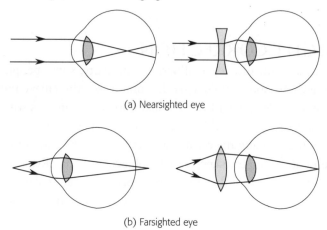

(a) Nearsighted eye

(b) Farsighted eye

Source: *Physics*, 3rd Ed, Giancoli, D.C., 1991, p. 658. © Reprinted by permission of Pearson Education Inc., Upper Saddle River, N.J.

In summary, the key idea here is that of refraction. Because light waves change direction as they slow down, they can be brought together at a focus to form a clear image. Any alterations in the curvature of the cornea, or loss of the ability of the lens to accommodate, cause the image to come to a focus either in front of or behind the retina. These condition are correctable with glasses or contact lenses.

10.4 ELECTROMAGNETIC WAVES

The range of frequencies of electromagnetic waves that are now used by the medical team has grown enormously in the last few years. At first, of course, nurses were limited to what their own eyes could tell them. Today, we use the whole range from the low frequency infrared waves for heat treatment to very high frequency X-rays to see into the body itself. Let's look at some of these, starting from the low frequency end of the spectrum.

Infrared waves

The infrared part of the spectrum covers a range from about 10^{11} Hz to 4×10^{15} Hz. These frequencies are felt by us as heat. The energy of the waves is absorbed near the surface of the body, converted to heat and thereby warms the skin. This has some value in relieving the pain of muscle and joint inflammation, if they happen to lie close to the surface of the skin.

Since we are warm-blooded mammals, with a body temperature of about 37°C, we are a source of these waves, beaming them out in all directions like a black light globe. These waves can then be detected by photographic film which is sensitive to infrared waves. Photographs of objects just below the surface are then possible and we can follow the progress of healing, or of events within the depths of the skin. The brighter the image on the detector, the hotter the area. Such techniques are referred to as **infrared thermography**. They can be used to detect tumours, since these areas are generally 1–2°C hotter than surrounding tissue. Some breast cancers can be detected in their earliest stages by this method. The detectors can show the image directly on photographic film, or convert it into electrical signals for display on a monitor.

Infrared thermography
A diagnostic technique which forms images of structures using the infrared radiation they emit.

Visible light

Visible light has frequencies higher than those of infrared waves. The human body is completely opaque to these frequencies, except of course the transparent materials of the eyeball. Besides the information gained by a visual examination of a client, there are numerous specialised uses for electromagnetic waves in this range.

Lasers

The word 'laser' stands for light amplification by stimulated emission of radiation. Ordinary visible light consists of waves of varying wavelengths, travelling in all directions from the source. Laser light is formed in such a way that it is coherent. That is, the light waves are of only one wavelength and they are almost exactly parallel to one another. As a result, laser light travels as a beam which can be very accurately focused. The most common medical uses thus far have been in ophthalmology, where the energy of the beam is used to treat haemorrhages in the retina. Referred to as photocoagulation, the heat energy coagulates the blood. In other cases, the beam is used to treat tears in the retina, or even its complete detachment, by creating local adhesions between the retina and surrounding tissue.

Endoscopes

These instruments are used to inspect body cavities. They are commonly long, semi-rigid tubes equipped with lenses and a light source. They can be inserted into a body cavity and used to examine and collect specimens. There are several types: bronchoscopes, for examining the lungs and bronchi; gastroscopes, for examining the gastrointestinal tract; and cystoscopes, for examining the bladder.

Fibre optics

These take advantage of a new technology that allows light to be transmitted down the length of a thin fibre. The fibre may be twisted and contorted without affecting the light in any way. These optical fibres, when bundled together, can be used in the same way as an endoscope. The advantage is their flexibility and smaller size. They can be used for surgical procedures deep within the body, reducing the need for large incisions.

Ultraviolet light (UV)

As we move up the electromagnetic spectrum, wavelengths get shorter and the frequencies get higher. Both the energy needed to produce the waves and the energy they carry get larger. Ultraviolet waves range from about 7.5×10^{15} to 7.5×10^{16} Hz. They are easily absorbed by the skin and do not penetrate past about 1 mm. Because their energy is rather higher than visible light, they can destroy or damage living cells. As a result, they are important factors in causing skin cancers. (A fuller discussion of radiation and cancer is given in Chapter 12.) Therefore they are not often used in either therapy or diagnosis. One exception, however, is for the treatment of psoriasis. This often chronic condition is characterised by over-rapid mitosis of skin cells, resulting in thickened layers of skin. Certain drugs such as methoxsalen are used to inhibit DNA duplication and these drugs are activated by bursts of UV waves.

X-ray radiation

Note that as the frequency goes up the word 'wave' is replaced by the word 'radiation' (light waves; X-ray radiation). In part, this recognises the increase in the energy needed to produce these waves; and it is also true that these high-energy waves act less and less like water waves and more and more like particles. As we'll see when discussing nuclear radiation in Chapter 12, the term 'radiation' is often used to refer to the particles that are emitted from the nucleus of the atom. An X-ray can be considered to be a high-energy particle of light. However, whether considered as wave or radiation, the role of transmitting energy from one place to another remains crucial.

X-ray frequencies range from about 3×10^{16} to 3×10^{19} Hz. They are produced by the simple, but atomically violent procedure of slamming fast-moving electrons into metal targets. Very high electrical potential energies of up to 2×10^5 V are required to get the electrons moving fast enough, so specialised equipment is required to produce X-rays. Their energy is so high that they can almost pass right through us. It is only when they collide with an atom or molecule within us that they are absorbed, and such collisions are rare. In other words, healthy tissue is transparent to X-rays, except bone which is so much more dense than other tissues. Other dense areas—abnormalities such as tumours, areas of necrosis, or foreign objects—also show up as darker patches on the X-ray film.

X-rays now have so many applications in nursing that it's impossible to say everything about them here. There are dangers associated with their use, discussed in Chapter 12. Here are a few highlights that reveal the importance of this diagnostic and therapeutic tool. As we'll see, there are three main ways of taking a radiograph, or X-ray:

1. In the first method, used in chest X-rays and in the examination of broken limbs, a burst of X-rays is sent through the body onto film lying behind the section of interest. This film is then examined by the medical team. A version of this technique is known as **fluoroscopy**. A continuous stream of X-rays is sent through the client and a moving picture of what is happening in the body can be seen.

2. In the second technique, a dye that is opaque to X-rays is first injected into the part of the body to be examined. This allows a picture to be taken of an area normally transparent to X-rays. Two examples of this are bronchography, where the bronchioles and lungs are dyed, and pulmonary angiography, where the heart and blood vessels are examined. Both X-ray film and fluoroscopy can be used.

3. The third method of taking an X-ray is referred to as **tomography**, from the Greek word for 'slice'. Here, the direction of the X-ray beam keeps changing, passing through the body from different angles. A thin beam of X-rays is passed through the tissue thousands of times from different directions, always passing through a common meeting-point. The resulting picture is a slice or cross-section of the tissue at that meeting-point. It is a two-dimensional slice, rather than a flat picture, but can be enhanced by

Fluoroscopy
A diagnostic technique using a continuous stream of X-rays to produce a moving picture of structures in the body.

Tomography
A diagnostic technique using X-rays, ultrasound, or radio frequency electromagnetic waves, generated from a moving source; a three-dimensional image can be formed by using computer technology.

a computer to three dimensions. The X-ray machine as well as the picture is controlled by the computer and therefore the technique used to be referred to as computer assisted tomography, or CAT for short. This has been shortened further to CT. For example, some of you may have heard of thoracic computerised tomography. It can see objects 100 times smaller than older X-ray techniques; down to about 1 mm.

10.5 THE EAR AND SOUND WAVES

Mrs D. also has some hearing loss. In particular, she is losing her ability to hear high frequency, low amplitude sounds clearly. What does this mean?

Sound is described with the same vocabulary as light waves and water waves. What we call audible sound is the origin, transmission and detection by the ear of vibrations, usually in the air. In the case of water waves, drops of water move up and down in response to the energy creating the wave. With sound waves, molecules of air move back and forth in response to the source of the sound. But as with all waves, the movement of the air molecules transmits the energy of the source of the sound. Figure 10.11 shows the pattern of air around a sound source.

Areas where the molecules are pushed close together are called areas of compression. Where the molecules are further apart than average, we have areas of rarefaction. Any source of sound creates these areas of compression and

FIGURE 10.11 Sound production.
Sound waves generated by a tuning fork travel through the air as pressure waves. The frequencies are measured in hertz (Hz), or the number of waves per second.

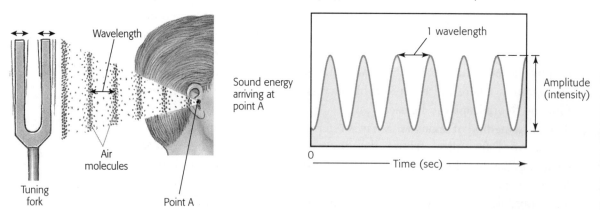

rarefaction, and their pattern determines the type of sound we hear. The source of the sound is vibrating (e.g. human vocal cords) and pushes against the surrounding air molecules. These bunch up into the areas of compression every time the source vibrates. The size of these areas of compression is the amplitude of the sound wave. The air molecules collide with one another, causing the next section of air to form similar areas of compression. And so the sound wave spreads out from the source, with the surrounding air vibrating at exactly the same rate as the source. Note that the molecules of air are not moving from the source to you, but only vibrating backwards and forwards. It is the energy of the source that is moving out from the source.

More specifically, for sound waves we have the following:

1. The amplitude of a sound wave is an indication of the amount of energy that produced the sound. It determines how loud the sound is, as measured in decibels. This is not the same thing as how loud the sound appears to us, because there are many frequencies which we cannot hear regardless of how much energy they have. Audible sound is only a small part of the sound spectrum.
2. The frequency of a sound wave is the number of areas of compression that strike the ear in a second (hertz) and we detect this as pitch; high-pitched noises (high frequency) versus low-pitched noises (low frequency).
3. The wavelength is the distance from one area of compression to another and it too is detected as pitch.

The speed of sound is not constant, but varies with whatever material the sound travels in and is usually faster in denser materials because the vibrations are more easily transmitted.

Let's look at these in a little more detail, with reference to Mrs D.

First, the relationship between frequency, wavelength and pitch. The human ear is sensitive to only a limited range of frequencies, from about 10 Hz to 20 000 Hz. But this sensitivity can change with age. As we get older, our ability to detect sounds of high frequency diminishes. Mrs D., like many people in their late 60s, has suffered a 15% drop in the range of frequencies she can hear easily. That is, any high-pitched sound, over about 17 000 Hz, will be difficult to hear. Look at the diagram of the ear in Figure 10.12.

It seems that, with age, the hair cells in the cochlea that respond to high frequency sounds degenerate. For this reason, it is a permanent condition, though a hearing-aid may alleviate it. The hearing-aid amplifies the sound and

FIGURE 10.12 The organ of Corti.

(a) A three-dimensional section of the cochlea, showing the compartments, tectorial membrane, and organ of Corti. (b) Diagrammatic and sectional views of the receptor hair cell complex of the organ of Corti.

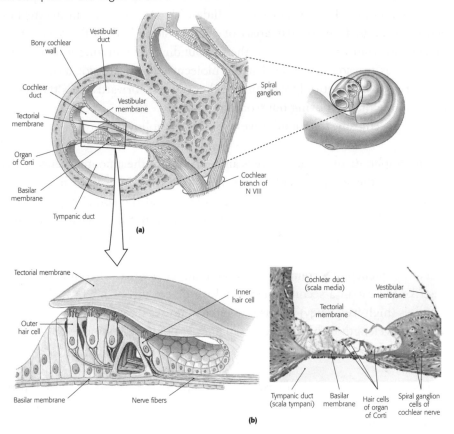

(a)

(b)

Source: *Fundamentals of Anatomy and Physiology*, 5th Ed, Martini et al, 2001, p. 566. © Reprinted by permission of Pearson Education Inc., Upper Saddle River, N.J.

transmits it directly to the auditory nerve through the surrounding bone. Low frequency sounds are naturally conducted through bone, and are therefore less affected. Of course, disease or accident could affect these too.

The usual speed of sound in air is about 344 m/s. A sound of only 1 Hz comes from a source that is vibrating once a second and is inaudible. In the 1 second between vibrations, the sound energy, of course, travels 344 metres. At a frequency of 5000 Hz, the sound travels 34 400 cm/5000 = 6.88 cm between vibrations and at 20 000 Hz the distance is down to 1.72 cm. These distances are the wavelengths of those particular sounds.

The amplitude or loudness of a sound is particularly interesting. Loudness depends, on one hand, on how much power, or energy/time, goes into producing the sound. This power is measured in decibels, discussed later in this chapter. But on the other hand, loudness also depends on how sensitive the ear is to that sound. For example, it makes no difference to us how much energy a sound of 50 000 Hz carries, because our ears are not built to detect it. As Figure 10.13 indicates, the human ear has different sensitivities to different frequencies.

So, even with the same energy, the ear is more responsive to a sound of frequency 15 000 Hz than to one of 4000 Hz, perhaps because human voices usually range around the upper end of the audible sound spectrum. For Mrs D., this may mean some inconvenience in hearing the higher voices of her grandchildren, the whistling of the kettle, or the upper notes of her favourite orchestral music. Loudness depends too, of course, on how close to the source of the sound we are. The nursing staff, as well as her grandchildren, may need to stand a little nearer to Mrs D. when talking to her.

Resonance

The physical principle that explains the ear's ability to detect sound is called **resonance**. It comes from the Latin word for 'sound'. Suppose a source of

Resonance
A physical property of some materials that allows them to vibrate at the same frequency as a source of sound with which they are in either direct or indirect contact.

FIGURE 10.13 Sensitivity of the human ear as a function of frequency.
Note that the frequency scale is 'logarithmic' to cover a wide range of frequencies.

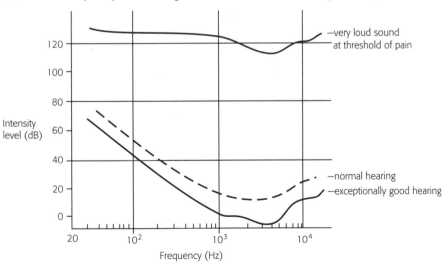

Source: *Physics*, 3rd Ed, Giancoli, D.C., 1991, p. 315. © Reprinted by permission of Pearson Education Inc., Upper Saddle River, N.J.

sound is vibrating at, say, 3000 Hz. A nearby object, or one actually touching the source, will then begin to vibrate at 3000 Hz too, if it can, because its molecules are being periodically struck by the molecules of the source of the sound. This ability to vibrate at the same frequency as a nearby source of sound is what we mean by resonance. The word 'nearby' means only that the sound must have enough energy left over after travelling to set the next object moving. A thunderstorm many kilometres distant is 'nearby' as far as resonance is concerned. The ear, as you can imagine, is a very good resonator. The eardrum, bones of the middle ear and the hair cells in the cochlea resonate at exactly the same frequency as the sounds they pick up. Schoolchildren often lay a wooden or plastic ruler on a table, with half of it sticking over the edge and then flick it so it vibrates up and down. The reason the sound is so loud is because the table resonates at the same frequency and adds its energy to the ruler's, thus amplifying it.

The ear also does a little amplifying of its own. From the time sound enters the ear until it is converted to electrical messages by the auditory nerve, the ear amplifies it about 180 times. This happens because the sound waves are channelled from a large area into a smaller one. For example, the size of the eardrum is much smaller than the opening of the ear and the energy is amplified 30 times. The small bone called the malleus that rests on the eardrum is much smaller than the eardrum and the amplification is three times. Finally, the cochlea adds an extra amplification of two times, for an overall amplification of 180 times.

Unfortunately, Mrs D. is showing the effects of ageing here too. The eardrum, for example, tends to become less flexible as we age. It therefore cannot resonate as easily and some of the amplification is lost. This explains her loss of amplitude problem, but many other factors can influence loudness. Conduction deafness results when the energy of the sound waves cannot travel all the way through the parts of the ear. Excess ear wax, a broken eardrum, middle ear infections, or damage to the bones of the middle ear are all possible causes. Damage to, or degeneration of, the hair cells and/or the auditory nerve causes perceptive deafness. This, remember, is Mrs D.'s other problem.

10.6 THE MEDICAL APPLICATIONS OF SOUND

Audible sound has been used for diagnosis for hundreds of years. Coughing, tapping the chest, thumping the abdomen—all have given the medical team some idea of the interior state of the body. Only recently, however, have we been able to make use of frequencies higher than the audible. The main technique uses **ultrasound** to gain information and treat certain conditions. Ultrasound frequencies start at about 20 000 Hz and can exceed 20 MHz.

The key to understanding ultrasound lies in recalling that the speed of sound is different in different materials. It is faster in dense material like bone, for example, than in less dense muscle tissue. Every time a sound wave passes from muscle to bone, the change of speed causes a reflection, or echo, from the point where the muscle and bone meet. The echo is then detected by a device like a microphone. A picture of the inside of the body can then be built up.

If we want to use ultrasound to see something small in detail, we need to use small wavelengths. A wavelength of 0.1 mm (15 MHz) lets us see objects that size. The catch is, the shorter the wavelength, the shorter the distance it can penetrate the body. To see further, longer wavelengths are required, but we lose detail. To see each part of the body with clarity, a characteristic wavelength and frequency must be used. Let's look at two examples of ultrasound in diagnosis.

Ultrasound
Sound waves of frequencies higher than those detectable by the human ear; i.e. over 20 000 Hz.

Ocular ultrasonography (up to 15 MHz)

This uses ultrasound to examine, among other things, the interior of an eye that is clouded by a cataract. The reflections from the back of the eye are electronically converted into a series of dots, which then build up a picture that can be displayed either on an oscilloscope or a television screen. This technique can give valuable information about the curvature of the cornea and the exact size of the eye. It could be used in Mrs D.'s case, to determine if her glaucoma has led to any change in the size of her eyeball. It is also used to detect malignancies, foreign objects and detached retinas. The technique is non-invasive and can be done with the eye closed.

Pelvic ultrasonography (about 1–3MHz)

This is perhaps the best known use of ultrasound. It can be used to examine the developing foetus, detect tumours, measure the size of pelvic organs and check for multiple pregnancies. The ultrasound beam is often scanned across the pelvic area, back and forth, building up a two-dimensional image. The picture can be displayed on a television screen, or photographed, so that both still and moving objects can be examined. While examination of the foetus was once considered routine, the current advice is to test with ultrasound only when required. There are some hints that the heat generated by the ultrasonic vibrations may be harmful. Such heat may, however, be useful in the treatment of muscle inflammation and damage.

Lithotripsy

This involves one of the few uses of ultrasound in treatment. The high-frequency vibrations are used to shatter kidney stones, making them small enough to pass harmlessly into the urine and out of the body.

10.7 NOISE POLLUTION

Decibel
One-tenth of a bel, the unit used to compare the power of two sounds. It is also used to indicate loudness.

To determine whether noise can affect health, we need to know a little about how it is measured. As mentioned earlier, loudness is related to the power of the sound, which is measured in **decibels (dB)**. The decibel is one-tenth of a bel, which is the unit used in acoustics to compare the power of two sounds. Power is measured in watts (1 watt = 1 joule/second). The comparison is usually between the power of some noise (let's call it N_1) and the power of a noise just loud enough to hear (call it N_o), the threshold of hearing. Then in simplest terms, a bel is N_1/N_o and it is a number with no units. If we always compare a noise that we want to study with the threshold of hearing noise, we can get a scale measured in decibels; see Table 10.2.

Studies of the effects of loud sounds show that about 100 dB can

TABLE 10.2 TYPICAL dB LEVELS

Source	dB
Rustling leaves	20
Quiet room at midnight	32
Conversational speech	60
Busy restaurant	65
Ringing alarm clock at 1 m	75
Heavy traffic	92
Home lawn mower	98
Jet airliner, 300 m above	115

Note: This is a logarithmic scale; an increase of 20 dB is an increase of 100 in power. Also, an increase of 1 dB is just detectable by the human ear as an increase in loudness. An increase in 1 decibel represents an increase of intensity of 26%.

permanently damage hearing if the sound is continued for long periods. It appears that the hair cells thicken to reduce sensitivity to the noise and after awhile this thickening is irreversible. At about 140 dB there may be severe pain in the ear and at 160 dB the eardrum may burst.

Besides physical damage to the ear, there is some evidence that prolonged exposure to noise leads to other health risks. Noise, especially undesirable or excessive noise, can be regarded as a source of stress. The body tends to react to all stress in much the same way. The amount of adrenocorticotrophic hormone (ACTH) in the bloodstream increases. Through a negative feedback loop, this hormone stimulates the adrenal cortex and sets off a range of metabolic changes by the body that is sometimes called the fight or flight response. A sound level of 65 dB at 10 kHz has been known to cause a 53% increase in ACTH secretion. Fortunately, most of these changes are transient, unless the exposure is prolonged over many years.

Other results show increases in adrenal medulla secretions, decreases in peripheral circulation and changes in heart function. For example, workers in high noise environments show more dysrhythmias, tachycardia and extra systoles than the general population. Keep in mind that these symptoms are not definitive proof; they may arise from many contributing causes.

Psychological effects from excess noise are also common. Most of the research has dealt with the problems associated with loss of sleep. Irritability, stress, loss of concentration, and even delirium and paranoia are well-known results of sleep deprivation. The sound need not be loud enough to bring the sleeper fully awake. Noises as low as 30–40 dB bring about changes in a sleeper's EEG that are usually associated with waking.

Other established results of excess noise in the environment are tension, resentment and frustration. Annoying in themselves, such feelings may well affect physical well-being. Tension, for example, often causes muscle tightening. In turn, this leads to fatigue, lactic acid build-up, and the secretion of ACTH mentioned above. It is no surprise that many governments, both state and federal, have enacted legislation that attempts to control the amount of noise in the environment. This is most clearly seen in occupational health legislation and in the environmental impact studies of airport sitings. It may well prove to be true, however, that stress-related illnesses, due in part to noise, will absorb more and more of the health worker's time.

Questions

Level 1

1. What is meant by the electromagnetic spectrum?
2. What is a wave? Define its characteristics: amplitude, frequency and wavelength.
3. What does frequency correspond to for light waves? For sound waves? What is the range of frequencies human beings can perceive for light? For sound?
4. Give an account of refraction. How does it assist in the act of seeing?
5. List three types of X-ray pictures and examples of their use in diagnosis.
6. Give examples of the use of infrared and ultraviolet light for diagnosis and treatment. Why must we be careful in using ultraviolet rays?
7. What are the properties of a laser that make it medically useful?
8. Define resonance. What important role does it play in the detection of sound by the ear?
9. What is the relationship between frequency and pitch for sound waves?
10. Give two examples of the use of ultrasound in medical diagnosis. Explain in simple terms how it works.

Level 2

11. What is the physical difference between red and blue light?
12. How does astigmatism interfere with proper vision?
13. Why must the shape of the lens of the eye continually change?
14. Distinguish between myopia and presbyopia. What type of lens is used to correct these conditions? Why?
15. What is the difference between light and X-rays?
16. Why are colours less distinct when viewed in dim light?
17. What is the difference between conduction deafness and perceptive deafness?
18. What is the decibel? If two sounds differ in loudness by 40 dB, what is their difference in power?

Level 3

19. Discuss the dangers associated with high levels of noise and with continuous, low-level noise.
20. If a client's hearing by bone conduction was found to be nearly normal, but severe hearing loss of airborne sounds was indicated, what causes of hearing loss are likely?
21. Under what conditions can very intense sounds strike a normal ear but not be heard?

CALCULATION PROBLEM BASED ON THE CASE STUDY

Imagine that Janice D. was not compliant with her eye drops. As a result she develops ocular pain leading to migraine, nausea and vomiting. She is ordered prochlorperazine (Stemetil) in a single deep IM injection. The recommended dose is 0.25 mg/kg of body weight, and Janice weighs 70 kg. You have available to you ampoules that contain 12.5 mg/mL of prochlorperazine. Calculate the volume that should be administered to Janice D.

Answer:
Prescribed dose = 0.25 mg/kg × 70 kg = 17.5 mg.
Injected volume = 17.5 mg/12.5 mg × 1 mL = 1.4 mL.

11

THE CLINICS PART 2: METABOLIC DISORDERS

DNA AND RNA, PROTEIN SYNTHESIS AND METABOLIC DISORDERS

Chapter outline

INTRODUCTION

Not all ill-health comes from disease, injury or lifestyle, of course. It is unfortunately true that some of us are born with conditions that affect health. Some of these are inherited, such as thalassaemia or haemophilia. Some are caused by environmental influences on the developing embryo; thalidomide and lead are tragic examples of this. In this chapter we turn our attention to these so-called metabolic disorders. To properly educate and treat clients with these disorders, it's essential to understand the underlying causes. We'll also apply some of the basic chemistry from earlier chapters.

Case Study: Daniel H.

Daniel H. is a baby of 16 months. He suffers from phenylketonuria (PKU) and visits the clinic once a month with his parents. His weight, pattern of growth and general state of health are assessed. Questions about his diet, behaviour and future course of treatment are raised and discussed. The chances of another child in the family also inheriting the disease is, of course, a great concern to Daniel H.'s parents. Genetic counselling is also available at the clinic.

Clinical note: It is possible that Daniel H. was given a large dose of phenylalanine orally, and the rate at which it disappeared from his bloodstream was measured. Administration of oral doses to young infants can be stressful and traumatic for staff, parents and baby.

11.1 DNA AND RNA

The natural question of why children resemble their parents is a very old one, but only in the last 50 years or so has the answer started to become clear. Within the nucleus of every cell of our body (except the red blood cells) are long, spiral-shaped molecules called DNA. **DNA** stands for deoxyribonucleic acid and it is within this molecule that genetic information is stored. A DNA molecule has an interesting, fairly simple structure (part of which was given in Chapter 4); see Figure 11.1.

DNA

The abbreviation for deoxyribose nucleic acid. It is the unit of heredity, containing genes that code for all the proteins essential for life.

FIGURE 11.1 Nucleic acids: RNA and DNA.

Nucleic acids are long chains of nucleotides. Each molecule starts at the sugar of the first nucleotide and ends at the phosphate group of the last member of the chain.

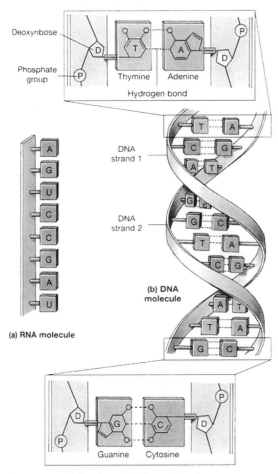

Source: *Fundamentals of Anatomy and Physiology*, 5th Ed, Martini et al, 2001, p. 57. © Reprinted by permission of Pearson Education Inc., Upper Saddle River, N.J

The deoxyribose part of the DNA is a type of sugar. (Remember, '-ose' means sweet.) As you might expect, it is a type of ribose that has fewer oxygen (deoxy) atoms than normal ribose. The term 'nucleic' refers to the fact that DNA is found in the nucleus of the cell. And the acid part of the DNA comes from phosphoric acid, H_3PO_4.

In Fig. 11.1 you will see four molecules labelled as complementary base pairs, or nitrogen bases. These are not the same, chemically, as the bases or alkalis discussed in Chapter 7. The four nitrogen bases found in DNA are cytosine, thymine, adenine and guanine. Their chemical structure is discussed later in this chapter.

Nucleotide

The building-block of the DNA molecule, it contains three parts: sugar, phosphate and one nitrogen base.

The combination of the three parts—sugar, acid and nitrogen base—is called a **nucleotide**. When the structure of DNA was first examined, the following picture of the molecule became clear. The sugar and the phosphoric acid parts form a chain or backbone from which the nitrogen bases hang:

Sugar—Acid—Sugar—Acid—Sugar—Acid—
| | |
base base base

The complete DNA molecule consists of two such backbones facing each other. They are held together by weak hydrogen bonds (Chapter 4) between the nitrogen bases:

Sugar — Acid — Sugar — Acid — Sugar — Acid—
| | |
adenine cytosine thymine
::: ::: ::: (hydrogen bonds)
thymine guanine adenine
| | |
Sugar — Acid — Sugar — Acid — Sugar — Acid—

There is one rule to remember about this. Adenine bonds only with thymine and cytosine bonds only with guanine. The hydrogen bonds between the nitrogen bases are so weak that it is easy to peel the two backbones apart, much like opening a zipper.

How does this molecule contain genetic information? The secret is in the sequence of the nitrogen bases. A series of three nucleotides on the one backbone is the code for a particular amino acid. For example, the series cytosine—cytosine—cytosine (ccc) means proline; and adenine—cytosine—guanine (acg) means threonine.

You may recall from Chapter 4 that only 20 amino acids are required to make all the proteins of the body. The four nitrogen bases, used in any combination of three, are more than enough to code for these 20 amino acids. Since a DNA molecule is millions of nucleotides long, there is enough room for all the information the cell needs to make any of the proteins required. It may now seem clear why any three nucleotides together are called a **codon**, as they code for an amino acid. Let's look at a larger piece of DNA and identify the codons and the amino acids they code for. Here, A = acid, S = sugar and the nitrogen bases are shown by their initials:

Codon
A combination of three nucleotide bases; a codon 'codes' for one specific amino acid.

S—A—S—A—S—A—S—A—S—A—S—A—S—A—S—A—S—A—S—A—S—
(DNA backbone)

| | | | | | | | | | |
A   A   A   C   T   C   C   A   C   T   G
_____   _____   _____ (nitrogen bases)
codon 1   codon 2   codon 3   (codons)

As a result of the processes of transcription and translation, to be discussed shortly, we find that:

codon 1 AAA stands for phenylalanine
codon 2 CTC stands for glutamic acid
codon 3 CAC stands for valine

The protein made from these three amino acids would then have the sequence —phenylalanine—glutamic acid—valine— as part of its structure. Notice too that we need only one backbone of the DNA molecule, not both, for the codons.

This seems to have taken us rather far from Daniel H. In fact, Daniel H. inherited half of his DNA from his mother and the other half from his father. Somewhere back in the ancestry of both his parents the sequence of nitrogen bases in their DNA was changed. Such a change is known as a **mutation** and usually means trouble. Sometimes a nitrogen base is left out, sometimes the position of one or more is shifted. For example, the DNA nitrogen base sequence

Mutation
A change in the codon sequence in a DNA molecule. It alters in the amino acid sequence and therefore the proteins made.

—CCTAAAGCCTTCGAA—

codes for the amino acid sequence

CCT   AAA   GCC   TTC   GAA
glycine—phenylalanine—arginine—lysine—leucine

Now suppose by mutation the original nitrogen base sequence is altered to

—CCTAAA*CCTTCGAA—

The * means that the base g has been lost. The amino acid sequence is also altered, as the nitrogen base pairs shift to the left to fill the gap. It now becomes

CCT   AAA   CCT   TCG AA—
glycine—phenylalanine—glycine—cysteine—

There are three things to say about this:

1. The missing nitrogen base (G in this case) does not leave a hole in the DNA. The sequence of nitrogen bases just moves over to fill its spot.
2. Different codons may stand for the same amino acid. For example, both GGG and GGA stand for glycine.
3. The different amino acid sequence means that a different protein is formed. This may be an unimportant protein, which the body can manage without or obtain through the diet. The real danger is that the protein not being made may be an essential one, such as an enzyme, which the body cannot do without.

This is the situation with Daniel H. His DNA sequence can no longer code for an enzyme called phenylalanine hydroxylase. Without it he cannot break down (catabolise) a substance called phenylalanine, which consequently accumulates in his body and exerts a toxic effect, especially on the brain. As well, in a healthy body, phenylalanine is normally broken down to produce another chemical called tyrosine; this is used to make, among other things, adrenaline. If the phenylalanine is not being catabolised, there is no tyrosine and therefore no adrenaline.

Chromosomes

While this is not a textbook on genetics, it is important to say a little more here about inheritance in order to fully understand the genetic counselling aspect of our case study. Let's return to the fact that Daniel H. received half of his DNA from each parent. The DNA in the nucleus occurs in structures called chromosomes. Chromosomes come in pairs, one half of each pair from each parent. A human being has 46 chromosomes in each cell, in 23 pairs. Daniel's father contributed 23 chromosomes from his sperm cell, and his mother contributed 23 from her egg cell, to make 46 in all. The string of codons that make a particular protein is known as the **gene** for that protein. So we can say that a chromosome contains a number of genes, made of codons, which are made of DNA. Genes for a specific protein are almost always found on only one of the chromosomes. Since research has shown that there are about 100 000 different genes, each chromosome has 100 000/23, or over 4000 genes.

Since each of the chromosomes in the pair comes from a different parent, they may not be exactly the same. The mother might have brown eyes, for example, which means she has a chromosome with a gene that produces a protein that, along with other factors, makes the eyes brown. The father, however, may have blue eyes, and therefore a gene for a different protein, one that helps make eyes blue. In these cases, one of the genes is usually **dominant** over the other. In the case of eye colour, brown is dominant over blue and the child will have brown eyes. The gene for blue eyes is said to be **recessive**, because the dominant gene is stronger. When both genes are the same (i.e. both code for the same protein), the genes are said to be **homozygous** (*homo* comes from the Greek word for 'same'). However, if they are not identical and therefore code for different proteins, the pair is called **heterozygous**, (*hetero* meaning 'different').

Both Daniel H.'s parents were heterozygous when it came to the gene for making phenylalanine hydroxylase. That is, one of their genes coded for it, but the other gene in the pair was recessive. Fortunately for them, though, the good gene produced enough of the enzyme for them to live normal lives; the good gene is dominant. His parents, however, are carriers of PKU. When Daniel was conceived, he unfortunately received only the recessive gene from each parent. Both members of the pair were defective and so he produced no enzyme. Technically, this is referred to as homozygous recessive. This type of genetic inheritance is illustrated by Figure 11.2.

Genetic counselling for Daniel H.'s parents should include the likelihood of having another baby with PKU. Since the chance of getting a recessive gene from each parent is one in two, we can see from the pattern in Figure 11.2 that the probability of two defective genes is only one in four, or 25%. Each child conceived would have a one in four chance of having PKU. Even if the conceived child has three siblings without PKU, his or her chances are still one in four.

Gene
A string of codons that make a particular protein.

Dominant gene
The gene in a heterozygous pair whose character is present in the organism; i.e. brown eyes genes are dominant over blue eyes genes, so the individual has brown eyes.

Recessive gene
The gene in a heterozygous pair whose character is not expressed in the organism. In a homozygous recessive individual, however, that gene is expressed; i.e. an individual with *two* blue eyes genes will have blue eyes.

Homozygous
The condition where both genes for a given protein, one from each parent, are the same; i.e. both code for a protein responsible for blue eyes.

Heterozygous
The genes in the pair, one from each parent, are different; e.g. one codes for a blue eyes protein, one for a brown eyes protein.

FIGURE 11.2 Daniel H. and the inheritance of PKU

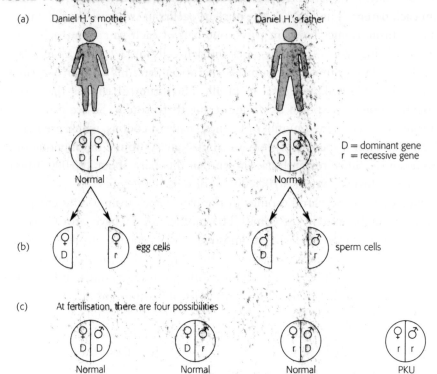

(a) Daniel H.'s mother   Daniel H.'s father

D = dominant gene
r = recessive gene

Normal   Normal

(b)   egg cells   sperm cells

(c) At fertilisation, there are four possibilities

Normal   Normal   Normal   PKU

11.2 PROTEIN SYNTHESIS

Just how, though, does the information in the codons get turned into proteins? If we think of DNA as a blueprint or recipe for making protein, then someone, or something, has to read the message and make the product. In our case, the reader of the DNA message is RNA.

RNA stands for ribonucleic acid. As you can see from the following diagram, it too is formed from a series of nucleotides. It's sugar, however, is a type with more oxygen in it than DNA's. Another difference between DNA and **RNA** is that thymine is not present in RNA. It is replaced by a similar nitrogen base called uracil. But as before, uracil binds only with adenine.

RNA
The abbreviation for ribose nucleic acid. Available in three forms, it translates the DNA code into protein at the ribosomes.

$$S - P - S - P - S - P - S - P - S - P - S \quad \text{(RNA backbone)}$$
$$\quad | \quad \quad | \quad \quad | \quad \quad | \quad \quad | \quad \quad |$$
$$U \quad G \quad A \quad A \quad U \quad C \quad \text{(nitrogen bases)}$$

DNA is found mostly in the nucleus of the cell. RNA has to leave the nucleus and go to the ribosomes, where the proteins are made. We label the RNA with names that tell us where it is and what it's doing. There are three types:

1. *Messenger RNA* (mRNA) is a working copy of the DNA in the nucleus. It's called 'messenger' because it carries the DNA message out of the nucleus to the ribosomes. It makes a copy of one backbone of the unzipped DNA in a process called transcription. The DNA's adenine, for example, matches up with a passing nucleotide of RNA which has uracil as its base. Guanine grabs a floating RNA nucleotide with cytosine as its base, and so on. In this way a length of mRNA, whose nitrogen base sequence matches the original DNA, is gradually built up. When the sequence is complete, it then detaches and floats off, leaving the nucleus and moving to a ribosome.
2. *Transfer RNA* (tRNA) has two ends. One end, called the anticodon, matches up to a mRNA codon. The other end holds the corresponding amino acid; see Figure 11.3. There is one kind of tRNA for each of the 20 amino acids.
3. *Ribosomal RNA* (rRNA) is found in the ribosomes. It holds mRNA and tRNA together.

The process of making a protein at the ribosome is known as translation. The easiest way to understand the process is to follow a diagram; see Figure 11.4.

The mRNA attaches to the small subunit of the ribosome, and the first codon is exposed. In the diagram, that codon is AUG. Along comes a tRNA with one end coding for UAC, which matches that on the mRNA. They thus bind together. (Remember, adenine binds with uracil and cytosine binds with guanine.) The other end of the tRNA holds the amino acid methionine, which is coded for by the AUG of the mRNA. Methionine is called an initiation codon, as it signals the start of the protein to be made. At this point the large subunit of the ribosome locks into place.

The mRNA now slides along the ribosome until the second codon, GUC, is exposed. The corresponding tRNA, with the matching CAG codon, binds to it. This tRNA carries the amino acid valine, whichis the one coded for by the mRNA's GUC. A chemical bond is formed between the methionine and the valine as they are now positioned side by side.

This process is repeated step by step, and this stage is referred to as elongation.

Eventually, the rRNA uncovers a mRNA codon that signals 'stop'. When that happens, the protein chain breaks free of the ribosome and is ready to be used by the cell. The process is summarised in Figure 11.5.

FIGURE 11.3 An overview of protein synthesis

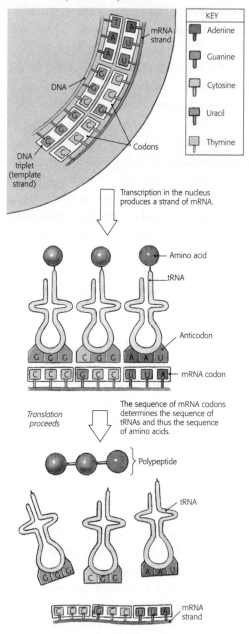

Source: *Fundamentals of Anatomy and Physiology*, 5th Ed, Martini et al, 2001, p. 93. © Reprinted by permission of Pearson Education Inc., Upper Saddle River, N.J.

FIGURE 11.4 Translation of DNA code into protein takes place at the ribosome, composed of a smaller and a larger subunit

(a) *Initiation*. The smaller ribosome subunit attaches to one end of a mRNA molecule. The first tRNA molecule, UAC, attaches to the AUG codon on the mRNA. The larger ribosome subunit then locks into place. (b) *Elongation*. The second tRNA molecule attaches to the appropriate site on the mRNA. A bond is formed between the two adjacent amino acids, and the first tRNA is released from the mRNA. A third tRNA then moves into place, and the process is repeated. (c) *Termination*. Finally the ribosome reaches a codon that signals 'stop'. The amino acids, now joined into a polypeptide or protein, are released, and the two ribosome subunits separate.

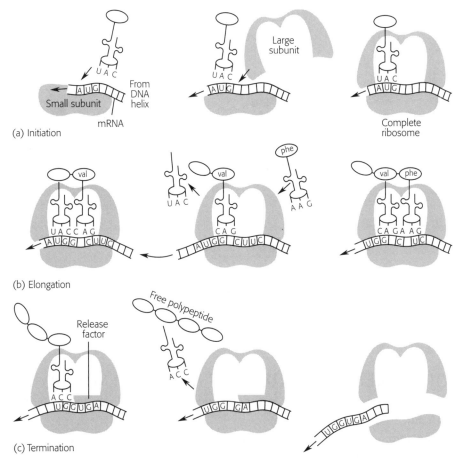

Source: *Invitation to Biology*, 4th Ed, Curtis, H. and Barnes, N.S., 1985, Worth Publishers, New York, p. 192. © 1972, 1977, 1981, 1985 by Worth Publishers. Used with permission.

FIGURE 11.5 Summary of protein synthesis

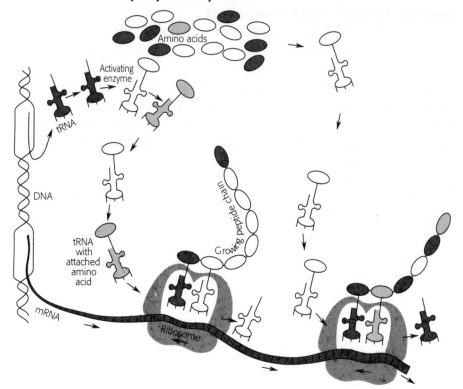

Source: *Invitation to Biology*, 4th Ed, Curtis, H. and Barnes, N.S., 1985, Worth Publishers, New York, p. 193. © 1972, 1977, 1981, 1985 by Worth Publishers. Used with permission.

As we can see, using Daniel H.'s case as an example, a change in the DNA sequence of his parents results in a change in the mRNA sequence, then in the tRNA sequence, and finally in the amino acid sequence.

SUPPLEMENT ON NUCLEIC ACID CHEMISTRY

Let' have a more detailed look at the chemical structure of the components of DNA and RNA; see Figure 11.6.

The joining of the deoxyribose to the phosphoric acid is a dehydration reaction; water is released. The next step, adding the nitrogen base to the sugar, is also a dehydration reaction. It results in the completion of the nucleotide.

The nucleotides are then linked one to another in the proper sequence to form the backbone of the DNA or RNA. With DNA, two strands then join

FIGURE 11.6 RNA and DNA.

(a) An RNA molecule consists of a single nucleotide chain. Its shape is determined by the sequence of nucleotides and the interactions between them. (b) A DNA molecule consists of a pair of nucleotide chains linked by hydrogen bonding between complementary base pairs.

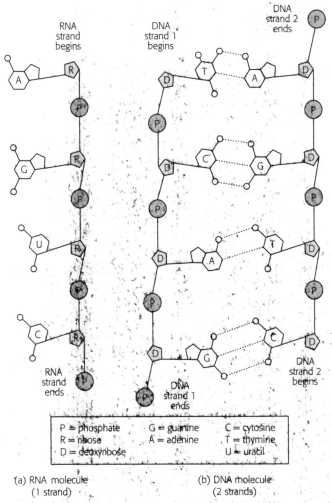

P = phosphate   G = guanine   C = cytosine
R = ribose   A = adenine   T = thymine
D = deoxyribose   U = uratil

(a) RNA molecule   (b) DNA molecule
    (1 strand)   (2 strands)

Source: *Fundamentals of Anatomy and Physiology*, 2nd Ed, Martini, F., 1992, p. 54. © Reprinted by permission of Pearson Education Inc., Upper Saddle River, N.J.

using hydrogen bonds and the whole long molecule is then twisted into a spiral, or helix. The term 'double helix' is a well-known description of the famous DNA molecule.

11.3 METABOLIC DYSFUNCTION

In one sense, we're already talking about metabolic dysfunction, because that's exactly what PKU is. There is some breakdown in the chemical metabolism of the body; some small disturbance in a complex network of chemical reactions. We looked at this, using a simple generalisation, in Chapter 5, in the case study of Rachel L. and digestion. Table 11.1 shows the general pattern associated with phenylketonuria (PKU).

Other examples of metabolic dysfunction include Tay-Sach disease, diabetes, and numerous abnormal haemoglobin conditions. One of these haemoglobin disorders is known as sickle cell anaemia. Here, the normal haemoglobin molecule differs in only one amino acid in the sequence; in that one place, glutamic acid has replaced valine.

Only some of these metabolic dysfunctions are inherited and some involve changes to more than one enzyme. Among the more familiar inherited conditions are Marfan's syndrome (which affected Abraham Lincoln and includes scoliosis as a symptom), albinism (caused by absence of tyrosine) and cleft palate. These last three do not show the simpler PKU pattern discussed above because alterations to more than one enzyme occur. For that reason, these conditions are usually referred to as polygenic.

In some cases, the trouble arises because one chemical in the body stops doing its job. This is sometimes quite normal and not inherited. As we age, for

TABLE 11.1 METABOLIC DYSFUNCTION INVOLVING PKU

Normal pattern	Dysfunction pattern
Dietary intake of phenylalanine	Dietary intake of phenylalanine
↓ O_2 + enzymes	⇸ O_2 + no enzyme
Tyrosine	Phenylalanine accumulation No tyrosine produced
↓ enzymes	⇸
Thyroid hormones: dopa, dopamine, adrenaline	No thyroid hormones produced

example, some hormones start to work and others shut down. With menopause, the levels of oestrogen circulating in the bloodstream tend to drop. Without oestrogen there to exert its homeostatic influence, other metabolic processes can get a little out of control. Osteoporosis is one such case. The main problem here is an increased amount of bone reabsorption. Oestrogen can, apparently, stimulate bone growth; when it disappears, unacceptably high rates of bone loss sometimes result.

11.4 TOXINS AND POLLUTANTS

The examples used so far in this chapter relate to metabolic dysfunctions which, except for osteoporosis, are inherited conditions. There are many of these, ranging from more easily managed ones such as diabetes, to the tragedies of gross malformations such as neurofibromatosis (the 'Elephant Man' condition).

However, not all metabolic disorders are familial, or inherited. Chemicals found in the environment can, once they enter the body, cause metabolic change and/or failure. A **toxin** is a chemical produced by a living organism that has a harmful effect on another living thing. Some snakes, ants and bees, certain plants, spiders—all these are familiar examples of organisms that produce toxins. A **pollutant**, on the other hand, is usually thought of as a by-product of human activity that can also harm other living things; for example, mercury, smog, insecticides, lead. There are some general ideas that apply to all pollutants and toxins, or xenobiotics as they are called collectively (*xeno* comes from the Greek word for 'foreign').

All xenobiotics must enter the body by moving through our natural barriers. There are four ways this can be done:

Toxin
A harmful xenobiotic that is the product of plant or animal metabolism.

Pollutant
A harmful xenobiotic that is either a naturally occurring inorganic chemical such as asbestos, or a product of human technology, such as DDT or petrol fumes.

1. *Ingestion*, via the gastrointestinal tract. For example, paint containing lead, or mercury in contaminated shellfish, may be eaten. What happens to it then depends on what sort of chemical it is and where or if it is absorbed into the body. Easily absorbed molecules are those that are lipid soluble (because they dissolve in the cell membrane); molecules not easily absorbed are ions (because they avoid electrostatic forces at the membrane pores) and those that are too large.

   If the xenobiotic is directly absorbed in the mouth, as can happen with some solvents, it will directly enter the general body circulation and bypass the liver. If, however, it is absorbed by the stomach or intestine, just as nutrients are, there is a better chance of it being metabolised by the liver into a harmless form and then excreted; more will be said about this later. The duodenum can also absorb some xenobiotics by pinocytosis (absorbtion by body cells).

2. *Inhalation*, via the respiratory tract. Examples of this are aerosols (which contain particles), gases and vapours. Let's look at them individually.

□ *Aerosol* particles, such as those of asbestos, are either swept away from the lungs by the cilia in the trachea, or manage to reach the bronchioles and alveoli, where they are engulfed by macrophages and carried to the lymphatic system. If they are toxic to the macrophages, then the latter will die and inflammation will result. Asbestos fibres damage the cells of the alveoli, leading to scarring and loss of lung function (asbestosis).

□ *Gases* such as carbon monoxide (CO) and hydrogen cyanide (HCN) have easy access to the body through the alveolar wall. Both of these bind strongly to haemoglobin, preventing it from carrying oxygen. Carbon monoxide poisoning is a recognised form of suicide. Other gases, such as sulphur dioxide (SO_2) and the oxides of nitrogen (NO, NO_2 and NO_3), dissolve in the fluids in the alveoli. They can form sulphuric and nitric acids, irritating and destroying pulmonary tissue.

□ *Vapours*, such as benzene, can also pass through the alveoli into the bloodstream, which means they can travel around the body to all major organs. Benzene, for example, leads to liver, bone marrow and kidney damage.

3. *Dermally*, through the skin. Some toxins, such as that found in poison ivy, can destroy the surface layers of the skin. Others increase the permeability of the skin, allowing xenobiotics easier passage. Still others, such as the insecticides using parathion, increase the ability of the xenobiotic to penetrate the skin. More will be said about these substances when we discuss organic pollutants further on.

4. By *injection*, through the skin. This is common in stings and bites by insects, snakes, spiders and other animals and plants.

Once inside the body, the xenobiotic usually will cross a cell membrane. This can be done in several ways. The foreign chemicals may be dissolved in water and pass through the pores in the membrane via diffusion (either passive or facilitated), by dissolving in the lipid layer of the membrane, by active transport, or by pinocytosis.

As a protective measure, the xenobiotics are sometimes stored around the body. Lipid-soluble compounds may be stored in what is referred to as 'depot fat', where they may remain for long periods. Bone is a storage site for some compounds, such as lead ions, Pb^{2+}, which replace the normal calcium. The lung can store small amounts of asbestos and there is some evidence that even plutonium can be stored there for short lengths of time. The liver stores cadmium and the kidneys can store small amounts of lead. Some heavy metals such as mercury can be bound into harmless forms by plasma proteins known as metallothionens. Hair and nails are very common storage areas.

The xenobiotics are, whenever possible, treated by the body and excreted. First of all, they may be directly excreted through the urine, faeces, sweat and the breath. In this case, they have remained unchanged. They may also be converted into harmless compounds called metabolites; such compounds are attacked by enzymes in the body and either anabolised or catabolised. Usually this changes them into water-soluble compounds, which are more easily excreted. This is an important function of the liver and, to a lesser extent, of the kidneys. A variation of this is to metabolise the xenobiotic into another, still actively toxic substance, but one that can then be more easily changed into a harmless one.

Each xenobiotic, whether toxin or pollutant, affects the body in its own way, so it is impossible to discuss them all. However, there are some common factors that influence the effect any xenobiotic can have, as follows:

- *Type of compound.* Some xenobiotics come in several forms, such as solid or liquid or gas; some come in different chemical compounds, such as liquid mercury or mercury salts. These different types have different effects.
- *Size of dose.* This is measured in units appropriate to the type of xenobiotic (e.g., gases may be measure in mg per cubic metre of air) and to its suspected concentration (e.g., in ppm, ppb, or g/L).
- *Method of entry into body.* This can affect how quickly a xenobiotic is stored or metabolised.
- *Duration of exposure.* Long-term exposure (chronic) may lead to an accumulation of damage which finally becomes detectable. Acute exposure sometimes causes rapid fatality, or sickness followed by complete recovery.
- *Synergistic factors.* Synergism refers to the process by which two factors working together have an effect far greater than that obtained by the two acting independently. For example, workers in the asbestos industry who also smoke have up to 90 times greater risk of lung cancer than non-smokers in the industry.
- *Host factors.* These are the individual's own characteristics of age, fitness, immunities and so on.

The following examples are representative of the more common xenobiotics and give us some idea of how these substances interfere with normal metabolism.

Heavy metals

Lead

Lead is a dense metal, found as a pollutant in the form of vehicle exhaust gases (inhaled as tetraethyl lead), in the soil as a pest control element and as an insecticide (ingested as lead arsenate). It has several well-known harmful effects, depending on whether the poisoning is acute or chronic. In cases of acute intake, usually of lead salts, the symptoms include burning pain in the gastrointestinal tract, bloody diarrhoea, and failure of heart, kidneys and liver. Chronic (long-term) exposure can lead to anaemia, inadequate haemoglobin production, and damage to the central nervous system.

Lead poisoning used to occur through the ingestion of paint containing lead (e.g. on children's cots), until these paints were banned. However, some older structures coated with such paint still pose a danger to workers who have to dismantle them with welding equipment. Organic lead in petrol also affects the brain and central nervous system by dissolving the nerve membranes and slowing the rate of electrical conduction.

To relate this more closely to nursing practice, consider the problem with haemoglobin production at the ribosome, which follows the pattern outlined earlier in this chapter. When haemoglobin is not being produced due to the inactivation of an important enzyme, two substrate molecules accumulate and appear in the urine. These two (d-amino laevulinic acid and coproporphyrin III) can be tested for clinically and indicate lead poisoning. The lack of product, haemoglobin, leads to anaemia.

Mercury

The World Health Organisation has set the safe level of mercury in the body at 0.05 ppm. Samples of human blood frequently show levels of 1–50 ppb, which are safe, but human hair may contain up to 5 ppm. This is an example of the storage of a toxic heavy metal in an inert form.

Mercury has had a long association with human beings. It was well known in ancient times, partly because it was used in the purification of gold metal. It was also once used in the production of felt hats. Workers in that industry ingested the mercury and eventually became insane; hence the expression 'mad as a hatter'. When the vapour is inhaled, it leads to gingivitis and proteinuria, though complete recovery is common. More dangerous are the inorganic salts of mercury, such as mercuric chloride. This has been used by suicides in the past. Only 1–2 grams can lead to severe vomiting, diarrhoea, bleeding from the gastrointestinal tract and profound kidney damage. Organic mercury compounds used in pesticides have been found in shellfish. These attack the central nervous system and can lead to permanent damage.

Arsenic

Safe levels of arsenic in workroom air have been set at 0.25 mg/m^3 per day, or 0.001 ppm. It has been calculated that human beings in settled areas may ingest up to 0.4 mg of arsenic a day in drinking water. This accumulates in the body, as follows:

Brain:	0.01 ppm
Blood:	0.04 ppm
Nails:	0.28 ppm
Hair:	0.46 ppm

Arsenic is the classic poison of murder mysteries. It is fairly easily disguised in milk, tea and porridge. In earlier times, arsenic was obtained by soaking it out of commercial flypapers. Today, it is available in certain weedkillers. The symptoms of arsenic ingestion are constriction of the throat and difficulty in swallowing, followed by severe vomiting and diarrhoea, acute shock and lowered pulse, which are sometimes fatal. Chronic doses, of course, were administered by the poisoners. The symptoms were a general feeling of malaise and vague pains in the abdomen. Usually only after death could a careful autopsy reveal the presence of accumulated arsenic.

Insecticides and fungicides

Insecticides come in basically two types: organochlorines and organophosphates. Both groups affect the nervous system.

Organophosphates

The organophosphates are organic molecules that contain phosphorus. Examples are parathion, malathion, and the trade names Guthion and Systox.

Organophosphates are characterised by short persistence and high toxicity. Persistence refers to the time they remain in the environment before they are metabolised into a different form, so these chemicals are dangerous for only a short length of time. They are absorbed through skin, inhaled and sometimes swallowed.

Organophosphates affect the body by inactivating the enzyme acetylcholine-sterase. This enzyme breaks down acetylcholine, which carries the nerve impulse from one nerve to another. The acetylcholine thus persists within the synapse. The symptoms are muscular twitching, weakness, tremors and convulsions. Recovery is often complete, within 3–4 months after exposure.

Organochlorines

These organic molecules contain chlorine and include DDT and dieldrin, along with lindane, chlordane and aldrin. They can attach to the nerve membrane and deactivate it by letting the sodium and potassium ions leak out. Organochlorines also have a long persistence. Dieldrin, for example, can remain unchanged in the soil for up to 4 years and in the body for over a year. For this reason, they accumulate in the food chain. They are also lipid-soluble.

Organochlorines enter the body by ingestion (e.g. eating food containing traces of insecticide), inhalation and dermally. They generally attack the central nervous system, making neurones depolarise randomly; symptoms include twitching and tremors. There is evidence linking them to cancer and to embryo abnormalities.

Snake venom

The most common enzyme found in snake venoms is lecithinase. It causes breakdown of the mitochondria, which stops the electron transport chain. As well, it seems capable of causing the release of acetylcholine at the nerve endings. Other enzymes present in smaller amounts inhibit the functions of ATP and NAD. Other effects include blood clotting, haemolysis, lowering of blood pressure and respiratory failure.

Poisonous spiders

The venom of the red-back spider, for example, is a neurotoxin. It attacks the junction between a nerve and a muscle, causing paralysis. This is often followed by weak pulse, cold skin, difficulty in breathing and delirium. Psychological symptoms that sometimes occur are intense anxiety and fear.

Smog

The word 'smog' is an amalgam of 'smoke' and 'fog', and is a combination of gases and particulate matter. It is a common air pollutant in industrialised cities. Though we briefly mentioned some of these gases earlier when discussing inhalation, let's look at them in greater detail.

Oxides of nitrogen

Nitric oxide is NO, formed by the combustion of the nitrogen in the air in automobile engines:

$$N_2 + O_2 \rightarrow 2NO$$

It has a **residence time** of 7–70 days. Residence time is the equivalent of **persistence** for gases; the amount of time they remain unchanged in the atmosphere.

Nitrogen dioxide, NO_2, is also formed from nitrous oxide and oxygen during combustion:

$$2NO + O_2 \rightarrow 2NO_2$$

It's a brownish gas, with a residence time of about 3 days. These two gases lead to pulmonary damage, the extent of which depends on the dosage:

- 1.6–5 ppm per hour: increased airway resistance
- 25–100 ppm per hour: acute, reversible bronchitis and pneumonitis
- > 100 ppm per hour: death from pulmonary oedema

Nitrogen dioxide also dissolves in the water in the atmosphere to form nitric acid, HNO_3. When inhaled, this is a pulmonary irritant.

Sulphur dioxide

Sulphur dioxide is SO_2 and is formed when fuels such as coal are burned to produce electricity or in steel mills. The gas has a residence time of about 4 days.

Persistence/Residence time

Indicators of how long a particular toxin or pollutant remains in the environment (or in the body) before being broken down to a harmless chemical. Residence time usually refers only to gases or vapours in the atmosphere.

In that time it too can dissolve in atmospheric water to produce sulphuric acid. Concentrations of sulphur dioxide in the air have the following effects:

- 8–10 ppm per hour: throat irritation
- > 20 ppm per hour: coughing

When SO_2 dissolves in the mucous membranes it leads to bronchitis and, if acute, to pulmonary oedema.

Ozone

The combustion of petrol also produces a variety of oxygen known as ozone (O_3), though traces of this gas exist normally in the atmosphere at all times. Recently natural ozone has become well known for the fact that it is disappearing, destroyed by chemical reaction with chlorofluorocarbons (CFCs) released from spray cans. Ozone provides protection from the sun's ultraviolet radiation. Without it, the incidence of skin cancer among humans is projected to rise substantially. However, O_3 has detrimental health benefits when inhaled. In particular, when mixed with the hydrocarbon compounds and nitric oxide emitted from car exhausts, it forms an eye irritant that is largely responsible for the weeping, sore eyes so characteristic of smog.

In general terms, air pollution leads to what is referred to as 'toxic insult' to the lungs. This results in one or more of the following conditions:

1. Cell death, or necrosis. This may lead to scarring and the formation of fibres on the pulmonary tissues.
2. Leakage of fluid or blood into lung space, or oedema. This interferes with the free exchange of gases in the alveoli.
3. Infiltration by inflammatory cells (inflammation).
4. Abnormal production of mucus. This increases expectoration, puts strain on the mucus-cleaning cilia cells and can block gas exchange.
5. Narrowing of the small airways, reducing ventilation.
6. Thickening of alveoli walls, impairing gas exchange.
7. Enlarging of alveoli, leading to emphysema.

As can be seen, all of these conditions, even if not present at an acute level, can contribute to the problems of anyone already suffering from a chronic obstructive airways disease.

While we may seem a long way from Daniel H.'s metabolic disorder, there are several connections. Pollutants present a medical problem precisely because they interfere with normal metabolic function. In that respect, they do not differ from PKU or diabetes. Second, the key to determining their presence and effects is careful clinical testing. That is, we need to determine the presence and amount of, say, asbestos in the lung just as we must determine the presence or lack of phenylalinase. Third, treatment is sometimes as much a matter of behavioural change as it is of medical intervention. For PKU, it's a matter of diet; for organochlorine poisoning, it concerns lifestyle and proper precautions. Here, the educational role of the nurse is felt very strongly. Finally, as many of the major diseases become less common due to immunisation and lifestyle improvement, health-care workers will have to become more alert and responsive to metabolic conditions that result from either inherited abnormalities or a damaged environment.

11.5 PRINCIPLES OF CLINICAL MEASUREMENT AND TESTING

Nurses spend a great deal of time monitoring their patient's condition. The medical staff needs access to the information that the body is using to control homeostasis, so samples of blood, urine (if possible), lymph and saliva are taken regularly. The samples are then checked for the type and amount of chemicals present. This chapter concludes with a brief discussion of these clinical measurements. Not only are they valuable diagnostic aids, but nurses are now given greater responsibility for carrying them out and correctly interpreting the results.

The presence or absence of particular chemicals is usually determined by their reaction with other chemicals. The reactions produce a colour change, or increase in heat, or a certain colour of light, which can be detected and measured. The technology of these detectors is becoming more complex and automatic. More important for our purposes is some understanding of the results of these tests, rather than the physics and chemistry behind them.

Homeostatic mechanisms in the body try to keep the level of all molecules within a safe range of values. Table 11.2 gives the normal amounts of some common substances in the blood or plasma.

Any significant deviations from these normal values usually indicate metabolic disorder, unless the client is on a course of therapy that deliberately upsets the normal pattern. Since these normal values are the ones we compare our clients' values with, it is also usual to refer to them as our standards. As well

TABLE 11.2 SAMPLE NORMAL VALUES

Substance	Approx. adult values	
Bilirubin	less than 17	μmol/L
Calcium	2.25	mmol/L
Cholesterol	5.2	mmol/L
Electrolytes		
sodium	140	mmol/L
potassium	4	mmol/L
bicarbonate	24	mmol/L
chloride	100	mmol/L
Glucose (fasting)	5.3	mmol/L
Iron		
males	21.5	μmol/L
females	14.3	μmol/L
Proteins (total)	60	g/L
Urea	3.3	mmol/L

as normal values, then, it's useful to have some idea of the range of acceptable values. After all, individuals vary slightly in their metabolisms and do not have values exactly the same as those on Table 11.2. Ranges are normally expressed as ± values, read 'plus or minus'. For example, urea: 3.3 ± 0.3 mmol/L means the value can range from 3.3 + 0.3 = 3.6 mmol/L to 3.3 − 0.3 = 3.0 mmol/L. We could write that as urea: 3.6–3.0 mmol/L; both expressions mean the same thing.

Ranges can also be written as ± a percentage—for example, proteins: 60 ± 2% g/L. The normal amount of protein can then vary from:

60 + 2% of 60 = 61.2 g/L to 60 − 2% of 60 = 58.8 g/L.

This could also be written as protein: 61.2–58.8 g/L.

Having both standards and acceptable ranges, the only thing left to check is our accuracy. This means two things here. First, how accurate is the technique by which we make the measurement? For example, all acids have the property of changing the colour of certain chemicals, which are called indicators for that reason. The most common of these indicators is litmus, which changes from blue to red when an acid is present. Litmus is useful for telling you that an acid is present, but not very accurate if you need to know how much acid is present (the pH). For that we need a pH meter, which can have an accuracy of ± 0.1%. This means that if the actual pH is 6.00, the pH meter will register between 5.94 and 6.06.

The second part of accuracy is how good we are at taking the sample in the first place. Nursing texts take a great deal of care to inform nurses of the correct collection techniques for urine, faeces, saliva, etc. Usually the samples must be of the right amount, from the correct area of the body, kept at the right temperature, not over-exposed to the air and used before too much time has passed. If these precautions are not taken, accuracy may be totally lost and the client's readings meaningless.

Now let's apply this to our case study. Daniel H. was diagnosed at birth by a simple, routine test. The higher than normal levels of phenylalanine in his urine were the key. Such clinical testing is becoming simpler and more inexpensive all the time. The number of things that can be tested for is also getting larger. The skills of nursing now include a whole range of testing and specimen collection techniques unavailable even a few years ago. In hospital settings the chemical analysis and interpretation are often done by specialists. However, there is growing demand for the health worker to operate and interpret the results of the testing instruments now becoming available to clients—for example, the home glucose level testing machines that diabetics routinely use.

There are several important things to keep in mind about clinical testing. First of all, as we said above, whoever does the testing must know if the result is within the normal range. Some of the substances in the blood commonly tested for, and the range of normal values, are shown in Table 11.3.

As well as knowing the range of normal values for the tests done, the nurse should be able to make some reasonable judgements about the usefulness of the reading. For example, how accurate is it? That means, how confident can we be that the answer is correct? Some measuring devices, like clinistix, can give only a rough idea of the amounts of the chemical tested for. For example, the clinistix test for protein is limited to telling us if there is little, some, or lots of protein in the urine. By contrast, some electronic testers can give us the answer in parts per million (ppm).

The other side of accuracy is when the machine gives you the right answer, but for the wrong reasons. Perhaps the client is taking medication that contains large amounts of the substance being tested for; or the collecting jars are not sterile enough; or the time between collection and testing is too great. If so, it is possible to get a **false positive result**. The client seems to have the chemicals in their body, but in fact they do not. Of course, these same

False positive result
An error in clinical measurement that implies certain chemicals are present in the sample when they in fact are not, or are there for reasons that have not been taken into account
(e.g. medication).

TABLE 11.3 CLINICAL MEASUREMENTS: BLOOD SAMPLES

Test	Conventional values	SI Units
Aldosterone	1–21 ng/dL	0.14–0.8 nmol/L
Amylase	56–190 IU/L	25–125 U/L
Bicarbonate	22–26 mEq/L	22–26 mmol/L
Calcium (total)	9.0–10.5 mg/dL	2.25–2.75 mmol/L
CO_2 (blood)	23–30 mEq/L	21–30 mmol/L
Cholesterol	150–250 mg/dL	3.90–6.50 mmol/L
GPUT	18.5–28.5 U/G haemoglobin	–
Glucagon	50–200 pg/mL	14–56 pmol/L
Glucose (fasting)	70–115 mg/dL	3.89–6.38 mmol/L
Insulin	4–20 µU/ml	36–179 pmol/L
Lead	120 µg/dl or less	< 1.0 µmol/L
Prolactin	2–15 ng/mL	2–15 µg/L
Protein (total)	6–8 g/dL	55–80 g/L
Testosterone	300–1200 ng/dL (men)	10–42 nmol/L
	30–95 ng/dL (women)	1.1–3.3 nmol/L
Uric acid	2.1–8.5 mg/dL (men)	0.14–0.48 mmol/L
	2.0–6.6 mg/dL (women)	0.09–0.36 mmol/L
Vitamin C	0.6–1.6 mg/dL	23–57 µmol/L

Note: The units in which results are given are osmoles, molar solutions, and milliequivalents (see Chapter 7).

conditions can also lead to a false negative reading, when the testing shows no trace of a chemical that is in fact present. This is why it's crucial to follow the testing procedures as closely as possible, including getting all relevant information from the client.

Questions

Level 1

1. Carefully define what is meant by a metabolic disorder. In what ways can it interfere with normal metabolic activity?
2. What specifically are the main problems with PKU?
3. What are the differences between the three types of RNA? What does each type do?
4. Distinguish between a nucleotide, a codon and a gene.
5. What is the difference between a toxin and a pollutant?
6. Define smog and list its common components and origin.

Level 2

7. Why is DNA said to be a molecular blueprint? How does it code for a protein?
8. Why is a mutation potentially harmful?
9. Distinguish between a false positive and a false negative result.
10. Outline the steps in translation (the making of a protein molecule) at the ribosome.
11. How do xenobiotics enter the body? How do they enter a cell? What types of molecules find it easiest to enter a cell? Why?
12. What are the main ways by which the body attempts to deal with a dangerous xenobiotic?
13. What are the main differences between organochlorines and organophosphates?
14. Referring to the three heavy metals discussed in this chapter, describe the common features of storage and symptoms.
15. Why do you think vomiting and diarrhoea are such common reactions to the entry of a xenobiotic into the gastrointestinal tract?

Level 3

16. Explain why Daniel H. had to get a defective gene from both parents to develop PKU, using the terms 'dominant', 'recessive', 'homozygous' and 'heterozygous'.
17. If Daniel H. has a new baby sister what are the chances that she will also have phenylketonuria?
18. What considerations should feature in genetic counselling given to Daniel H.'s parents?
19. Why does changing the sequence of bases in a strand of DNA lead to incorrect protein production?

CALCULATION PROBLEM BASED ON THE CASE STUDY

Daniel H. is having neonatal apnoea (difficulty in breathing) after all the testing. He weighs 3.5 kg, and is ordered theophylline 12 hourly. The syrup available in the nursery contains 80 mg of theophylline in 15 mL. The loading dose (or stat dose) is 6 mg/kg and the maintenance dose is 2 mg/kg every 12 hours. Calculate the number of millilitres of syrup required for (a) the stat dose, and (b) the maintenance dose.

Answer:
(a) Stat dose: = 6 mg/kg × 3.5 kg = 21 mg.
Stat volume = 21 mg/80 mg ×15 mL = 3.9 mL syrup.
(b) Maintenance dose 2 mg/kg × 3.5 kg = 7 mg.
Maintenance volume = 7 mg/80 mg × 15 mL = 1.3 mL syrup.

THE ONCOLOGY WARD AND CLIENT CARE

NUCLEAR RADIATION AND ITS EFFECTS ON THE BODY

Chapter outline

12.1 **The structure of the atomic nucleus.** Your objectives: to be able to describe the structure of the nucleus; to understand the forces that operate there; and to be able to define the terms 'fission' and 'isotope'.

12.2 **Nuclear radiation: types, units, half-life and dosages.** Your objectives: to distinguish between the types of radiation used in clinical settings; to be able to use radiation units and dosages correctly; and to be able to define half-life.

12.3 **The effects of radiation on living tissue.** Your objective: to understand the positive and negative effects of radiation in a clinical setting.

12.4 **Radiation treatment and diagnosis.** Your objective: to be able to discuss the place of radiation in diagnosis and treatment.

INTRODUCTION

There is no doubt that radioactive materials are increasingly important in health assessment and treatment. The diagnosis of many types of dysfunction and the treatment of clients suffering from various cancers is well known. While the number of people suffering from nuclear radiation injury in Australia is low, it's still necessary for the nurse to understand the harmful effects of such radiation on cells and tissues. In this chapter we look closely at the physics of nuclear radiation, its uses in the hospital, and its effects on health. As well, some attention will be given to the longer-lasting genetic effects of radiation.

Case Study: Mr James H.

Mr H. is one of the large number of elderly males suffering from prostatic cancer. It is the most common type of cancer occurring in men, and has a mortality rate of about 30%. Mr H. is 59 years old, and has been treated on and off for 3 months. At the moment, he is in the oncology ward to have radioactive iodine-125 (I-125) beads implanted into his prostate gland. These will remain there permanently, although their level of radioactivity will decrease over time. This treatment is usually performed only when the cancer is an advanced one. The nurses on the ward are responsible for explaining to him the need for temporary isolation, the nature of the treatment and the possible side-effects. They must also be able to protect themselves from the implanted radiation.

12.1 THE STRUCTURE OF THE ATOMIC NUCLEUS

Radioactivity was first discovered and studied by Becquerel at the turn of the century. Since then, we have come to understand something about the various types of radiation and their effects upon living organisms. However, there is still a great deal of uncertainty about exactly how radiation affects living cells. Therefore, some of what will be discussed in this chapter is still speculative. As its name implies, nuclear radiation originates in the nucleus of the atom; Figure 12.1 shows the structure of a carbon nucleus.

Protons and neutrons are collectively known as nucleons, because they are found in the nucleus. All neutrons, as mentioned in Chapter 4, are electrically neutral, while protons are positively charged and therefore repel one another. How, then, is it possible for these protons to remain together in the nucleus of an atom? The answer lies in the interaction of the protons and neutrons in what is called the **strong nuclear force**. This force is called 'strong' because it can overcome the electrostatic force of repulsion between the protons. When the energy associated with this strong force is released from the nucleus in an uncontrolled way, it provides the tremendous energy that appears in nuclear weapons. When under our control, it is the source of energy in nuclear power stations. In small, carefully controlled amounts it is of great use in the hospital.

Normally, atoms are quite stable and the particles in the nuclei stick together permanently. However, there are several situations in which a nucleus is not stable, and its energy is released. Since there are different types of instability of the nucleus, it is not surprising that the released energy comes in several forms.

First, the atom may have so many protons that even large numbers of neutrons cannot prevent them from flying apart. For example, uranium has 92 protons in its nucleus, but even with 146 neutrons the repelling force is too great for complete stability. The uranium nucleus spontaneously splits into two

Strong nuclear force
One of the four fundamental forces of nature, it holds the protons within the nucleus, opposing their electrostatic repulsion.

FIGURE 12.1 A carbon atom nucleus with six protons and six neutrons.

This is known as carbon-12, or $^{12}_{6}C$.

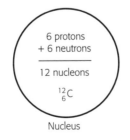

Nucleus

pieces, the protons repel each other and the two new nuclei travel apart at tremendous speeds, crashing into nearby molecules and occasionally damaging them. This splitting of uranium is known as **nuclear fission**. It occurs naturally in the uranium found in the earth and we take advantage of it in nuclear-power generating stations.

A second way in which a nucleus may be unstable occurs when there are too many neutrons relative to the number of protons. Take another look at the carbon atom, in Figure 12.2.

Note that it comes in several different varieties: one with six neutrons (shown in Fig 12.1), one with seven neutrons, and one with eight neutrons. Each variety is known as an **isotope**. Most of the carbon in the world is of the isotope known as carbon-12. That is, 6 protons + 6 neutrons = 12 nucleons. This is sometimes written as $^{12}_{6}C$. The subscript 6 (called the atomic number) shows the number of protons in the nucleus and the superscript 12 (called the mass number) shows the number of protons and neutrons in the nucleus. The other two isotopes would therefore be written as $^{13}_{6}C$ and $^{14}_{6}C$.

It turns out that the nucleus of $^{14}_{6}C$ is unstable. This time, however, the carbon nucleus does not split into pieces. Instead, one of the neutrons changes into a proton and an electron:

$$neutron \rightarrow proton + electron$$

Neutrons are normally held together by a fourth force, called the **weak nuclear force**. Under unstable conditions, however, it cannot prevent the neutron from breaking down.

The result of all this is as follows:

1. The number of protons in the nucleus goes up from six to seven, changing the atom from carbon to nitrogen.

Fission
The splitting of the nucleus of a radioactive isotope into two or more fragments, along with the release of a great deal of energy.

Isotope
A variety of atomic nucleus. The isotopes of an element all have the same number of protons, but differ in the number of neutrons present.

Weak nuclear force
This fundamental force is responsible for holding the neutron together. Outside the nucleus of an atom, a neutron spontaneously splits into a proton and an electron.

FIGURE 12.2 The two isotopes of carbon: ^{13}C and ^{14}C

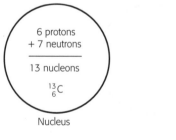

6 protons
+ 7 neutrons
———
13 nucleons

$^{13}_{6}C$

Nucleus

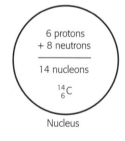

6 protons
+ 8 neutrons
———
14 nucleons

$^{14}_{6}C$

Nucleus

2. The number of neutrons in the nucleus goes down from eight to seven, so the symbol is now $^{14}_{7}N$.
3. The electron goes flying out of the nucleus at enormous speed; in fact, it is now one type of nuclear radiation.

The general point here is that some atoms have unstable isotopes. Such isotopes are called **radioisotopes**, because they emit radiation when they break down, radiation that comes flying out at high speed. This breakdown is referred to as nuclear decay. This type of radiation is therefore known as **nuclear radiation** and the isotope is said to be radioactive. In some cases, as with uranium, the energy of these smaller particles is used to turn water into steam to generate electricity. More important here, these particles can also be used to kill cancer cells, or help in diagnosis.

A third type of instability arises when the nucleus contains a surplus of energy, usually because the nucleus was formed as a result of the fission of some other nucleus. When this happens, the unstable nucleus sheds its excess energy in another form of nuclear radiation to be described below.

Radioisotope
An isotope that is radioactive; it emits radiation from an unstable nucleus.

Nuclear radiation
The particles, including gamma photons, that are emitted from the nucleus of a radioisotope. There are three types:
alpha rays;
the nucleus of a helium atom
beta rays;
an electron
gamma rays;
high energy photons of light

12.2 NUCLEAR RADIATION: TYPES, UNITS, HALF-LIFE AND DOSAGES

Even though there are many different radioisotopes, there are only three types of nuclear radiation. They have come to be known as alpha (α), beta (β) and gamma (γ) rays.

Alpha rays

Alpha rays have the same structure as the nucleus of a helium atom; that is, two protons and two neutrons, $^{4}_{2}He$. Being relatively large particles, they usually only travel through a few centimetres of air before they run into nearby molecules. So they are stopped by an ordinary sheet of paper, for example, or light clothing. An example of alpha decay is the breakdown of radium:

$$^{226}_{82}Ra \rightarrow {}^{222}_{86}Rn \text{ (radon)} + {}^{4}_{2}He \text{ (}\alpha\text{ ray)}$$

Alpha rays are not normally a danger to us if they come from the outside, as they cannot penetrate the skin (though they may damage skin cells). If a source of alpha rays is taken into the body, however, they can damage nearby cells.

Beta rays

Beta rays are the electrons referred to above as coming from the decay of a neutron in the nucleus. They penetrate many metres of air and even thin layers of metal before they are stopped. An example is the decay of radioactive phosphorus:

$$^{32}_{15}P \rightarrow {}^{32}_{16}S + \text{electron } (\beta \text{ ray})$$

Again, because these are easily shielded, they are not usually a grave danger from the outside, but once within the body they can cause cellular injury.

Gamma rays

Gamma rays are high-energy light waves. In Chapter 10 we mentioned that such high-energy electromagnetic waves act somewhat like particles. The name given to such particles of light, you may remember, is photons. Gamma ray photons behave in some ways as if they were particles like alpha and beta radiation and this will be discussed later in this chapter. If we think of them as waves, they have frequencies comparable to those of powerful X-rays, well over 10^{17} Hz. The only difference between a gamma photon and an X-ray is that the gamma radiation comes from the nucleus of the atom. And, like X-rays, they have no mass and no electric charge. They are very penetrating, passing easily through 2 kilometres of air or 30 centimetres of lead. As mentioned earlier, these are the ones given off by an unstable nucleus that has a surplus of energy. A common medical isotope that shows this is barium:

$$^{137}_{56}Ba \text{ (unstable)} \rightarrow {}^{137}_{56}Ba \text{ (stable)} + \gamma$$

Radiation therapy that uses an external source of radiation commonly uses gamma rays, as they can easily penetrate the skin. When ingested, both beta and gamma radiation are extremely dangerous. In the case of a nuclear accident involving sources of gamma rays, or any radioisotope in gaseous form, ingestion is a serious danger.

The units for measuring radiation

There are two ways of measuring radioactivity. First, we can count the number of particles (α, β and/or γ) emitted by the source. (For our purposes, consider the gamma photon as a particle.) Alternatively, we can measure the effect of those particles when they strike an object, such as ourselves.

In the first case, the original unit was known as the curie (Ci), named in honour of Marie Curie, who discovered both radium and polonium. In those units, 1 Ci was equal to the amount of radiation given off by 1 gram of radium, which is an amazing 3.7×10^{10} decays per second. That has turned out to be rather too large a unit for medical purposes, which uses doses in μCi sizes. The new SI unit is the **becquerel (Bq)**:

> 1 becquerel = 1 decay per second = 2.7×10^{-11} Ci

Second, we can measure the effect radiation has on the material it strikes. Since that material absorbs radiation and its energy, the old unit for this was called the rad, which stood for radiation absorbed dose. It has since been replaced by the SI unit called the **gray (Gy)**, where 1 Gy = 100 rad and 1 Gy = 1 J/kg. That is, 1 Gy is the amount of radiation that deposits 1 joule of energy to every kilogram of the absorbing material. Since bone, for example, is denser than muscle tissue, it absorbs more radiation, so a source of radiation deposits more energy in bone (which stops more of it passing through) than in muscle (which stops less).

Since different types of radiation cause different amounts of damage to tissue, we also need a unit that measures damage to tissue. This unit used to be the rem, which stands for rad equivalent man, but it has been replaced by the **sievert (Sv)** (1 Sv = 100 rem). In a sense, sieverts measure the effective dose of radiation—effective, that is, in the sense of damaging cells.

> effective dose (Sv) = dose (Gy) × QF

QF stands for 'quality factor'. It takes into account that alpha rays, for example, are 20 times more damaging to tissue than either beta or gamma. That is, the QF of alphas is 20, while the QF for both betas and gammas is 1.

What this means is that 1 Sv of any type of radiation does about the same amount of biological damage. For example, 100 Sv of alpha rays does the same damage as 100 Sv of beta rays. Keep in mind, though, that 100 Sv of alpha rays is only a dose of 5 Gy, while 100 Sv of beta rays is a dose of 100 Gy.

Becquerel (Bq)
The unit of radiation measuring the number of decays per second; 1 Bq = 1 decay/sec.

Gray (Gy)
The unit of radiation that measures how much energy is given to the material absorbing it.

Sievert (Sv)
The unit of radiation that measures the damage done by the absorbed dose.

Half-life

Since the radioactive nuclei keep emitting particles, after a while all the atoms in the radioactive isotope become stable and the emission of radiation stops. The time it takes for all the radioactive atoms to decay depends on the isotope of that element. For some, the nuclei are very unstable and break down in a matter of thousandths of a second, hours or days. For others, it's a matter of years or centuries.

To measure these different rates of decay, scientists use the idea of **radioactive half-life**. A half-life is the time it takes for one half of a specific amount of isotope to decay. For example, radioactive iodine-131, used in cancer therapy, has a half-life of 8.08 days. That means, if you start out with 2 grams of I-131, then 8.08 days later half of it will have decayed into stable atoms and you will have only 1 gram of I-131 left. After a further 8.08 days, only 0.5 gram of I-131 will remain; after another 8.08 days, 0.25 gram remains; and so on. Some medically important radioisotopes, types of decay and half-lives are shown in Table 12.1.

The nurse must have some idea of half-life for two reasons. First, it is necessary in some cases to isolate the client until the level of radiation in the body is low enough for them to be safely approached. The time this takes is related to the half-life. Second, if the radioisotope is in a liquid form, it will be excreted by the body as part of normal body metabolism. The nurse needs to know for how long the collection of urine, changing of dressings and washing will demand special safety precautions, which are discussed later in this chapter.

This second factor is important enough to be given its own name, the **biological half-life**. This is the time it takes for half of the radioisotope to be eliminated from the body by the normal processes of excretion. It bears no

Radioactive half-life, or Physical half-life
The time taken for one half of a given amount of a radioactive substance to decay into either stable or unstable forms.

Biological half-life
The time taken for one half of a given amount of a radioactive substance to be eliminated from the body by normal excretion.

TABLE 12.1 HALF-LIVES OF SELECTED RADIOISOTOPES

Element	Symbol of isotope	Type of radiation	Half-life
Phosphorus	P-32	β	14.28 days
Potassium	K-40	β and γ	1×10^9 years
Cobalt	Co-60	β and γ	5.271 years
Strontium	Sr-90	β	28.8 years
Iodine	I-131	β and γ	8.08 days
Gold	Au-198	β and γ	2.7 days
Radium	Ra-226	α and γ	1.6×10^3 years

relationship to the physical half-life, but must be taken into account when sources of radiation are placed in the body.

In a clinical setting, radiologists are most concerned with the effective half-life of the radioisotopes that are administered. This is a combination of the physical half-life, which indicates of how active the source is, and the biological half-life, which indicates how long it remains in the body. Mathematically, the effective half-life is calculated as follows:

$$\frac{1}{T_{eff}} = \frac{1}{T_{phs}} + \frac{1}{T_{biol}}$$

Using radioactive iodine as an example, we know that the physical half-life is 8.08 days. Assume the biological half-life is 3 days. Then the effective half-life will be:

$1/T_{eff} = 1/8.08 + 1/3 = 11.08/24.24$
$T_{eff} = 24.24/11.08 = 2.19$ days

12.3 THE EFFECTS OF RADIATION ON LIVING TISSUE

There are two main ways by which radiation can damage cells: directly and indirectly. Let's take them one at a time and start with the sketch of molecules in a cell being bombarded with alpha or beta rays, in Figure 12.3.

Here we can see the results of **direct radiation damage** to the molecules in the cell. The radiation physically strikes and breaks apart the molecules, which are referred to as 'targets'. There are two main theories about what is happening and why the damage is occurring.

The multiple-target, single-hit model suggests that each cell in the body has several vital targets, such as the DNA molecules. To kill the cell, all these targets

Direct radiation damage
Damage done by the physical impact of radiation on a molecule.

FIGURE 12.3 Direct radiation damage: an alpha particle strikes a large, complex molecule and damages it.

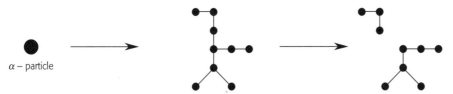

α – particle

must be hit by radiation within a short period, but they need only be hit once. Then, according to the model, the cell cannot repair itself and dies.

The repair model proposes that there are no vital targets. Instead, radiation damages cell components and this damage is then repaired by normal cell metabolism. The cell dies only when it runs out of the materials it needs to make repairs. Suppose they stop making spare parts for your lawnmower: it 'dies'.

Indirect radiation damage

Damage done to a cell by the ionisation reactions that occur due to the passage of ionising radiation through that cell.

Figure 12.4 illustrates what is meant by **indirect radiation damage** to the cell.

The particles that fly out from the nucleus have so much energy that they can strip the electrons off any atoms they are passing close to. (Nuclear radiation, therefore, is also referred to as ionising radiation. Another form of ionising radiation is X-rays.) The atoms left behind will then be ions. Some of these ions are highly reactive; that is, they chemically combine with other molecules quickly and strongly. As a result, these newly formed molecules may become useless, or even dangerous, to the cell. A very common example of this is water in the cell:

$$\text{H}_2\text{O} \quad \xrightarrow{\text{irradiated}} \quad \text{H}_2\text{O}^+ \quad + \quad \text{an electron}$$

FIGURE 12.4(a) Indirect radiation damage, caused by ionising radiation.
An alpha particle passes close by a water molecule and removes one of its electrons. The water molecule is now positively charged. $\text{H}_2\text{O} \rightarrow \text{H}_2\text{O}^+ + e^-$

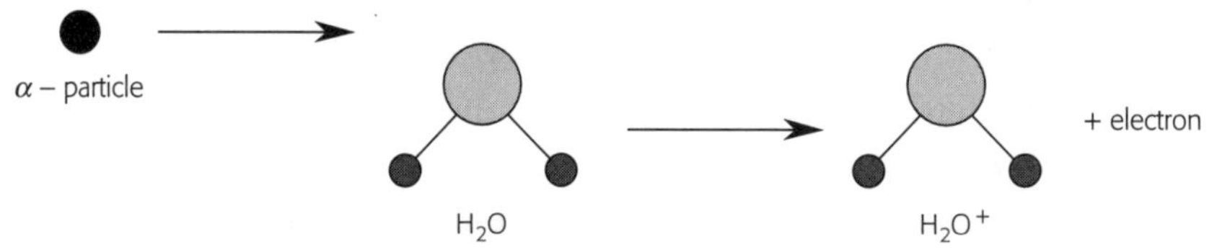

α – particle

H_2O

H_2O^+

+ electron

FIGURE 12.4(b) H_2O^+ is very reactive.
Here it combines with a complex protein molecule, destroying its structure, and therefore its usefulness.

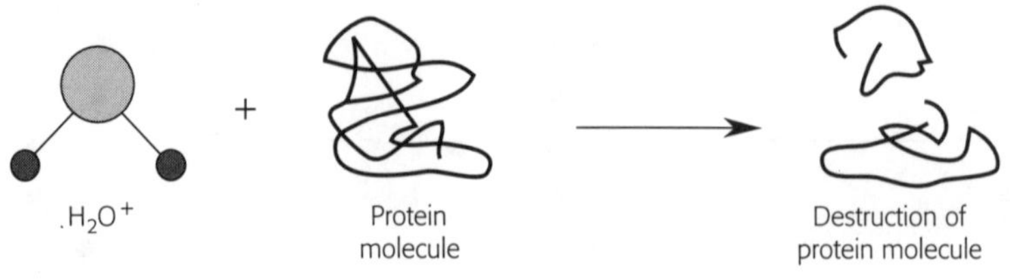

.H_2O^+

+

Protein molecule

Destruction of protein molecule

(*Note*: In the equation above, the electron is not a β ray, because it did not come from the nucleus of any atom in the water molecule.)

This ion form of water is extremely reactive. If it combines with a DNA molecule, for example, the DNA can no longer fulfil its normal function.

It might seem that the more ionisations that take place as radiation travels through a cell, the greater the chance of one of these damaging reactions taking place. However, this is complicated by the fact that, if there are many ionisations, the ions will more likely react with each other than with molecules in the cell. As well, the amount of water in the cell varies, making some cells more susceptible than others. This indirect damage partly explains the differences in QF (quality factor) mentioned earlier. Alpha particles, being so big and electrically charged, produce very high numbers of ionisations. Beta particles, charged but small, produce less, and gamma rays, because they are uncharged and mass-less, produce very few ionisations. Also, they cannot affect charged particles through the electrostatic force.

So far we've discussed radiation as though it was being used to kill a cell. This is certainly true if we want to destroy a tumour, for example. But another concern for nurses is the opposite of this, when radiation causes a cancer. In this case, the radiation damage is limited to the DNA in the chromosomes. The DNA strands (as discussed in Chapter 11) may be broken but later rejoin in the original pattern, in which case everything is fine. Or the DNA strands may break, then rejoin with a strand from a different DNA molecule. Under certain conditions still not understood, this may lead to uncontrolled cell division; that is, a cancer. Or the DNA strands may break and fail to rejoin at all.

This usually leads to cell death; see Figure 12.5.

One of the implications of this for normal cells is that cells that are actively dividing are most at risk from damage by radiation. Such cells include some of those in the skin, the lining of the gastrointestinal tract, the gonads, and the hair-forming cells. Radiation therapy often produces definite changes to these areas, even if they are not the ones being directly treated. As discussed later, such changes include burning, slowed cell division, or cell death. A further implication here is that an embryo is clearly a group of rapidly dividing cells. Radiation striking the embryo could lead to birth defects. For example, if those cells which develop into the liver were damaged, the foetus could be born with a liver reduced in size, or malformed. Sperm and egg cells, before fertilisation takes place, could also be affected. Their DNA could be damaged, leading to mutations as outlined in Chapter 11.

FIGURE 12.5 Potential radiation damage to a DNA molecule

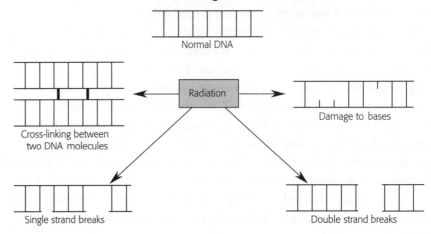

Lethal dosages

Whether we're talking of the harmful results or beneficial uses of ionising radiation, the effect of these rays is to damage or kill cells. If enough cells are destroyed, the whole organism may die. Such large amounts of radiation are referred to as **lethal doses**, abbreviated as LD. This dosage is usually expressed as $LD_{50}(30)$, which means the amount of radiation that will kill 50% of a population within 30 days. It's important to remember that this could mean a population of cells in a tumour, bacteria in an infected wound, or people in a city. The population must be specified clearly. Table 12.2 gives some typical $LD_{50}(30)$ values.

LD_{50} values for different times (e.g. 60 days rather than 30) are used not only for radiation, but for toxic chemicals and environmental pollutants. Since many lethal dose values cannot be determined for human beings, we are forced to rely on results from experimental animals such as rats. Care therefore must be taken in assuming that the figures given for people are exact.

Lethal dosage
The amount of radiation that will kill 50% of a population within a specified time.

TABLE 12.2 LD_{50} VALUES FOR SELECTED POPULATIONS

Population	$LD_{50}(30)$ value in grays
Dogs	2.44
Humans	4.5 (speculative)
Rats	7.96
Bacteria	1000
Viruses	10 000

What is cancer? The destruction of a tumour

Mr H. is hoping that the implanted radiation will destroy the cells of his cancer. But let's be clear about what exactly is meant by a cancer.

In most nursing texts, a cancer or carcinoma is also referred to as a neoplasia or neoplasm, meaning 'new formation'. It is also termed a 'malignant tumour' or 'malignant neoplasm'. In any case, the main features are as follows:

1. There is unrestrained mitotic division; that is, the cells keep multiplying without check.
2. There is a loss of cellular specialisation; that is, the cell no longer acts like a normal cell of its type (a liver cell, for example), but may take on new characteristics. The cells may also change in physical appearance.
3. The new cells produced are not effectively removed from the body by the normal processes of autolysis (self-breakdown) and/or phagocytosis (removal by the body's scavenger cells).

This third point is particularly important. One definition of a cancer is a group of cells that reproduce faster than they are removed. If the removal process was fast enough, the tumour could not grow large and threaten surrounding areas, which of course is the main danger from a cancer. It crowds out normal tissue and invades body cavities, cutting off supplies of oxygen and nutrients. Malignant tumour cells also show metastasis, another main danger; cells break away from the tumour and are carried to other areas of the body. There, they resume the process of unchecked growth and crowding.

To destroy a tumour, three changes in the tumour's cells must take place:

1. Mitosis must be stopped or slowed. This means that the radiation must either damage the DNA, or interfere with ATP production in the mitochondria. Because they are rapidly dividing, cancer cells are usually more susceptible to such damage. The graph in Figure 12.6 gives information on dosage and effects on dividing cells.
2. Any non-mitotic cells must be killed. It is the presence of such cells, remember, that squeezes out the healthy, surrounding cells.
3. Damaged and/or dead tumour cells must be removed from the body. If not, they form a lump of necrotic tissue which in itself is a health hazard.

Determining the correct dose to destroy a cancer without causing too much damage to surrounding tissue is not an easy task. Many factors must be kept in

FIGURE 12.6 The effect of increasing doses of radiation in living cells

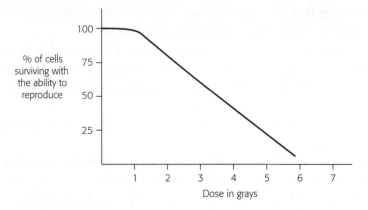

mind. For example, the dose must be matched to the size of the tumour; consider the graph in Figure 12.7.

As the tumour mass increases, the dosage requirement goes up. Again, there are several reasons for this. The cells in the middle of the tumour have less water and oxygen available to them. As we have seen, the H_2O molecules in a cell are easily turned into reactive ions by radiation. If these cells have fewer H_2O and O_2 molecules present, they are less vulnerable to indirect damage. As well, once the outer cells of the tumour are killed they form a barrier to the passage of further radiation; and the first dose tends to kill those cells that are, for one reason or another, the most sensitive to radiation. Those that are left are likely to be the most resistant.

FIGURE 12.7 The relationship between the mass of a tumour and the amount of radiation required to destroy it

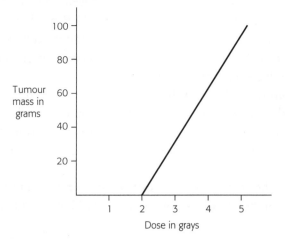

The dead cells must be removed from the body. So the normal course of treatment is to give a first dose to kill the sensitive, well-oxygenated cells. Then allow time for these dead cells to be removed and for the deeper cells to gain more water and oxygen. This time also allows normal cells the chance to repair themselves. The timing must be closely calculated, so that the dead cells are not replaced. Then give a second dose to kill the previously hypoxic (low in oxygen) cells. This process may have to be repeated several times.

12.4 RADIATION TREATMENT AND DIAGNOSIS

Returning to the case of Mr H., let's try to bring together some of the things mentioned so far in this chapter. His prostatic cancer is being treated with a radioactive isotope, I-125. This is a source of both beta and gamma rays. Being implanted in the prostate gland, the beta rays can easily strike the tumour cells. The gamma rays will pass through the tumour, hopefully doing some damage to it on the way and pass out of Mr H.'s body. Until enough time has passed, determined by the physical half-life of I-125, Mr. H. is therefore a source of gamma radiation and will need to spend some time in isolation.

Several aspects of this case need further elaboration. If Mr H. is in isolation, what about the effects on the nursing staff who treat him? There are three aspects that help to reduce the exposure of the nurse to the client's radiation:

1. Time of exposure—the shorter, the better. This calls for planned, short visits to the client and efficient carrying out of tasks.
2. Distance from client—the greater the better. In fact, this is an example of what is known as an inverse-square relationship. The effect of the exposure decreases as the square of the distance. Twice as far away means only one-quarter the exposure; three times as far away means only one-ninth the exposure. This rule suggests that, when possible, stand at the end of the bed furthest from the radioactive source.
3. Shielding—both thickness and type of material. Some materials are better at stopping nuclear radiation than others. Lead is a common type of shielding material. It may be advisable to wear protective clothing, or surround the treated site with some form of shielding.

Mr H. is receiving treatment with an implanted source. Here, the I-125 is sealed in metal beads. This is a specific example of what is generally known as internal **radiography**; see Table 12.3.

There are two types of internal radiography: sealed and unsealed. In sealed

Radiography
The diagnosis and/or treatment of a tumour by radiation. It may be internal (sealed or unsealed) or external.

TABLE 12.3 SAMPLE INTERNAL RADIOTHERAPY TREATMENTS

Isotope	Radiation	How it is given	Condition treated
P-32	β	Liquid: IV, orally & injected	Leukaemia and polycythaemia
Au-198*	β and γ	Liquid: injected	Lung, pleura & peritoneal cancer
I-131	β and γ	Liquid: orally	Thyroid carcinoma
Y-90	β and γ	Liquid: orally & injected	Rheumatoid arthritis
Cs-137	α and β	Sealed in beads	Gynaecological cancers
Ir-192	β	Sealed in wires	Oral cancers

*Au-198 is usually given to the terminally ill; clients who die soon after treatment must be identified as containing a radioactive substance.

treatment, the radioisotope is contained in a solid, usually a metal of some kind. It is then sewn or pinned into place and either left permanently or removed after a certain period. Radium-226 and cobalt-60 are commonly used sealed radioisotopes. The mouth, cervix and prostate are examples of areas treated in this way. These are cases where the nurse, knowing the site of implant, stands as far from it as possible when attending to the client. In the case of sealed sources, body wastes are not radioactive.

Unsealed internal radiotherapy uses liquid radioisotopes. These fluids are either drunk or injected, though they may be given intravenously. Common examples are I-131, P-32 and Au-198. If these are passed out of the body during excretion, the urine is often collected in lead-lined containers for the nurse's protection.

As well as internal radiotherapy, there is external radiotherapy. The source of the radiation is placed outside the client's body. A beam of gamma rays is directed onto the tumour, from either one or many directions. Of course, such radiation must pass through the intervening tissues, so the dosage must be carefully controlled to prevent damage to these healthy areas.

With external radiotherapy, the client should be told that there is no pain associated with the treatment, but there are often associated side-effects. Early reactions to the treatment, within 48 hours, may include erythema (redness) or blanching of the skin, leading to dry or moist desquamation (skin loss); and effects on the gastro-intestinal tract, such as nausea, vomiting and anorexia; see Table 12.4.

These are expected because, as was said earlier, cells that are rapidly dividing,

TABLE 12.4 SAMPLE DOSAGE/SYMPTOM RELATIONS

Dosage in grays	Symptoms
0.5–1	Lymphocytes decline
1.2	Anorexia
1.7	Nausea
2.1	Vomiting
2.4	Diarrhoea

such as skin cells and cells in the gut, are most at risk of radiation damage. Note that, contrary to some popular opinions, clients do not become radioactive themselves with external radiotherapy.

Late reactions to the treatment may include radionecrosis, which is the death of cells or tissues as a result of the therapy. Particularly at risk are the genitals, kidney and muscles. There may also be hair loss (either temporary or permanent, depending on the dosage), loss of sweat glands in the skin and changes in skin pigmentation. Even longer-lasting effects are lowered blood cell counts, increased risk of infection due to a lowered immune system, anaemia and blood-clotting problems.

The effects of radiation therapy on the body vary somewhat, depending on the client. For example, children are more at risk, because most of their cells are dividing as their body grows. The elderly are also at greater risk, because their immune systems are already lowered. There is some evidence that women are generally more at risk than men, though it is not clear why.

Not all cancers are equally sensitive to radiotherapy. For reasons that are not well understood, tumours differ in what is known as **radiosensitivity**. Highly sensitive tumours include leukaemia, Hodgkin's disease, seminoma of the testis and nephroblastoma. In these cases, radiation therapy can make a tumour completely disappear. Among the less sensitive, but still treatable, are carcinomas of the skin, breast, lung and ovary. Again, complete disappearance is possible but is not always expected. The poorly sensitive include carcinomas of the rectum and kidney and malignant melanomas. Only temporary restraint on the growth of the tumour is expected in these cases.

Radiosensitivity
Indicates the susceptibility to destruction by radiation of a particular type of tissue or tumour.

Radiation for diagnosis

As well as their uses in treatment, radioisotopes have an important role to play in diagnosis. Because they are radioactive, they can be detected in the body using various detectors. Probably the most familiar of these is known as a Geiger counter, though that is not its correct technical name. Geiger counters are the ones shown in the movies, detecting the presence of radioactivity by emitting a series of clicks. The detector is properly known as a Geiger–Muller tube. Whenever an alpha, beta or gamma ray passes through it, the tube detects any ionisation of the gas that fills the detector. A signal is then sent to a counter, which either gives off an audible click, or quietly counts the number of ionisations and displays them on a dial. Radiation can also be detected with photographic film. The badges worn by those who work in the proximity of radioactive materials is made of such film. How 'fogged' the film gets indicates how much radiation they've been exposed to.

Three common examples of the use of radioisotopes in diagnosis are given here. The nurse must become familiar with these techniques, to inform the client about treatment procedures and purposes. Whenever radiation is used there are also associated health risks to the client and health workers which should be understood.

1. The first of these diagnostic techniques is the use of radioisotopes to get a picture of internal structures. An example is a renal scan, or radionuclide renal imaging. The term 'radionuclide' refers to an isotope with a radioactive nucleus. In this case, the radioisotope is technetium-99, which is injected peripherally and travels quickly to the kidneys. A gamma ray detector is passed over the kidney area and records the image on either X-ray or Polaroid film. The technetium-99 is usually excreted within 24 hours. The resulting picture can be used to detect renal infarctions and atherosclerosis, monitor the progress of a transplanted kidney, detect some renal diseases such as glomerulonephritis and reveal abscesses and cysts. For these tests, the dosage is so low that no special precautions are necessary, though there is always a need for care when working with radioactive material.

2. A second diagnostic technique is to use radioisotopes to check on the proper functioning of an organ. A good example of this is the radioactive iodine uptake test. The health workers may need to know, for example, if the thyroid gland is functioning properly. If it is, it will trap and retain a certain amount of iodine in a given time, because the thyroid uses iodine to produce several of its hormones. A measured quantity of I-131 is drunk by the client and a gamma ray detector is placed by the throat. It determines the amount of I-131 that ends up in the thyroid in a given amount of time. This test allows the diagnosis of hyperthyroidism (caused by decreased thyroid

activity) and hypothyroidism (greater than normal thyroid activity). There are nursing-related procedures involved in this test, for example determining the client's iodine intake from diet and medication; dietary control before the test; timing of the administration of I-131 and the start of detection; and changes of medication before, during and after the test. Again, the levels of radiation are low and isolation of the client is not necessary.

3. A third diagnostic use is pregnancy testing. All such tests are based on the detection of a hormone known as human chorionic gonadotrophin (HCG), which is secreted after an ovum is fertilised. If HCG is present in a blood or urine sample, it can be detected by mixing it with a chemical substance called an antibody, which causes the two to clump together. There are two tests for HCG, based on radioisotopes. One is based on a technique which has many uses, called radioimmunoassay (RIA). In this application, it is a highly sensitive test that can be performed in 1–5 hours and can be performed very early after conception, even before the first missed menstrual period. The other is radioreceptor assay (RRA), which can be done in 1 hour and is very accurate.

In both cases, the sample from the client is mixed with a solution that contains the antibody and HCG that has been radioactively 'labelled'. This means that one of its elements has been replaced with a radioisotope of that element. For example, normal carbon-12 might be replaced with carbon-14. Any clumps that form will either be radioactive because of the labelled HCG, or normal from the client's HCG. The ratio of radioactive to normal clumps is evidence for the pregnancy. Both these tests are safe, as the client does not receive any dose of radiation, but they require the proper equipment and health workers must be experienced in their use.

Questions

Level 1

1. Define radioactivity. What are the characteristics of the radiations that are emitted?
2. Define the following units, distinguishing between becquerel, gray and sievert.
3. What does $LD_{50}(90)$ mean? Does it refer to any particular type of organism?
4. What factors characterise a cancer?
5. What general principles are involved in the destruction of a tumour by radiation?
6. What precautions can be taken to prevent unnecessary exposure when caring for a client undergoing internal radiography?
7. Define nuclear fission. Why is it a source of nuclear radiation?

Level 2

8. Distinguish between the strong and the weak nuclear forces. What role does each play in the emission of radiation?
9. Give an example of alpha, beta and gamma decay. Use both word and chemical symbol equations.
10. How does a radioisotope differ from an isotope?
11. What is meant by the term 'half-life'? Assuming the half-life of strontium-90 is 29 years, make a table that shows the amount of radioactive strontium left after 29, 58, 87, 116 and 145 years.
12. How does the amount of radiation administered during therapy relate to the effects on tumour cells? Mention tumour size in your answer too.
13. Distinguish between physical, biological and effective half-life. How are they related? If the physical half-life is 4 days and the biological half-life is 2 days, calculate the effective half-life.

Level 3

14. Why is there a quality factor in the relationship between sieverts and grays?
15. How does the biological half-life affect the health precautions taken by a nurse caring for a client undergoing unsealed internal radiography?
16. What is the difference between direct and indirect radiation damage? Outline clearly how each one causes harm.
17. Discuss three common uses of nuclear radiation for diagnosis. Which one poses the greatest risk for the client? Why?
18. Discuss what is meant by an inverse square law and relate it to nursing safety when dealing with radiation.

19. Look back at Chapter 10 and discuss the differences between X-ray radiation and nuclear radiation.

20. Why are radioimmunoassay and radioreceptorassay useful in the detection of pregnancy? How do they work?

SAMPLE DRUG CALCULATION PROBLEM RELATED TO THE CASE STUDY

Postoperatively, Mr H. is ordered 7.5 mg IM of morphine for pain relief prn. Ampoules of morphine sulphate contain 10 mg/mL of morphine. What volume should be withdrawn for each dose?

Answer:

Injection volume = 7.5 mg/10 mg × 1 mL = 0.75 mL of morphine sulphate.

APPENDIX 1
FUNDAMENTAL MATHEMATICS

PERCENTAGE, RATIO AND ALGEBRAIC EQUATIONS

One of the greatest concerns of nursing students can be the amount of mathematics in their courses. For those who are returning to study with a poor mathematics background, it is often a cause of anxiety. As well as a lack of confidence in their mathematical skills, some students have acquired a strong dislike of maths.

It would be wrong to pretend that this one small appendix could remove anyone's strong feelings about mathematics, or instantly provide them with the skills they need to succeed well in their studies. No one book can make you 'good' at maths unless you are prepared to spend some time working fairly hard at it. As with all things, it takes time and effort to become proficient at mathematical calculations. And you will not succeed unless you manage to convince yourself that you can; that mathematical skills are not beyond your ability.

Introductory science courses often demand three things of the students. First is the ability to do fundamental mathematical operations; percentage and ratio, for example. Second, they expect them to be able to interpret algebraic formulae; e.g., $CO = SV \times HR$ (cardiac output = stroke volume × heart rate). And third, they expect them to understand and use the metric system; e.g., 3 centimetres = 30 millimetres. We will look at the first two of these here, as the metric system was discussed in Chapter 1 of this text.

FUNDAMENTAL MATHEMATICAL OPERATIONS
Percentages

The words **per cent**, written as %, mean 'per 100'. This gives the number of items for every 100 possible. That is, 5% means '5 out of 100'; 15% means '15 out of 100'; and so on. A test score of 68% means, out of every 100 questions asked, 68 were answered correctly.

Percentages are easily calculated. Suppose a test has 25 questions; that's the number possible. Now imagine we got only 15 of those possible questions correct. To show this as a percentage:

$$\frac{\text{Number correct}}{\text{Number possible}} = \frac{15}{25} \times 100 = 60\%$$

The procedure is always the same.

Examples

1. What % of 75 is 6? 6/75 × 100 = 8%
2. What % is 18 out of 56? 18/56 × 100 = 32.1%
3. What is 5% of 35? Since 5% means 5 per 100, or 5/100, 5% × 35 = 5/100 × 35 = 1.75
4. What is 18% of 650? 18/100 × 650 = 117

Let's consider percentage from a different point of view, and use a practical example. Suppose that, as part of your science course, you run across a 5% solution of dextrose in a chemistry or nursing class (dextrose is a type of sugar used in IV solutions). In these cases, 5% means 5 grams of dextrose dissolved in 100 millilitres of water. Again, the 5% means 5 per 100; here, 5g/100 mL. Therefore, if we had a beaker with 200 mL of dextrose solution in it, we would know that 10 g of dextrose were dissolved in the 200 mL of water.

Ratios

You may sometimes see two items expressed as a **ratio** of each other. For example, the notes in your practical book might refer to a 1:1000 saline solution. That particular example means 1 g of salt dissolved in 1000 mL of water. As another example, say 1 tablet contains 200 mg of a given drug. The ratio of drug to tablet is written as drug:tablet, and is shown as 200 mg:1.

Ratios are simply a way of comparing two or more items. Imagine there are 36 people in your science class: 15 males and 21 females. The ratio of males to females is:

males:females = 15:21

Another way of thinking about this ratio is as follows. For this group of 36 people, 15 are males and 21 are females. If this ratio were true for *every* science class in the school, then we can say that, for *every* group of 36, 15 are male and 21 are female. Therefore, if there are five science classes, then the total number of students is $36 \times 5 = 180$; and

total number of males is $15 \times 5 = 75$
total number of females is $221 \times 5 = 105$

The ratio of males to females is both 15:21 and 75:105. These are referred to as **equivalent** ratios.

It is usual to express ratios using the *smallest possible equivalent ratio*. In the example above, the ratio of 15:21 can be expressed as the smaller 5:7, since both 15 and 21 can be evenly divided by 3. Similarly, a ratio if 30:9000 is more simply written as 1:300.

The reverse calculation is equally straightforward. Suppose you are told that the ratio of body fat between males and females is 1:2. That is:

body fat (males):body fat (females) = 1:2

This means that, **on average**, for every 1 gram of body fat a male has, a female will have 2 grams. So, if an average male of age 35 years has 1600 grams of body fat, a female of equivalent age will have $1600 \times 2 = 3200$ grams.

Another, more complicated example involves eye colour. Suppose we are told that, in Australia, the ratio of people with brown eyes to people with blue eyes is 4:1. That is, brown eyes:blue eyes = 4:1. Now suppose that the population of Australia includes 10 million people with either brown or blue eyes.

Then, for every five of these people (since $4 + 1 = 5$), there will be four people with brown eyes and one person with blue eyes.

10 000 000 people in groups of five = 10 000 000 ÷ 5 = 2 000 000 groups of five
2 000 000 × 4 = 8 000 000 people with brown eyes
2 000 000 × 1 = 2 000 000 people with blue eyes

A ratio can also be written as a fraction, as a decimal, or as a percentage.

As a fraction: 1:25 = 1/25
As a decimal: 1:25 = 1 ÷ 25 = 0.04
As a percent: 1:25 = 1/25 × 100 = 4%

READING ALGEBRAIC EQUATIONS

An algebraic equation is another way of showing relationships between two or more items. This is perhaps best shown with an example. My current bank balance is the sum of the money I have deposited minus the money I have withdrawn plus any interest I have earned. Written as an algebraic equation, it appears as:

Balance = Deposits − Withdrawals + Interest

I can replace the words above with appropriate symbols to make it shorter:

$B = D - W + I$

At the beginning of this Appendix there appeared the equation:

$CO = SV \times HR$

In words, this equation tells us that the **Cardiac Output** of the heart (that is, the volume of blood each ventricle pumps in a minute) is found by multiply the **Stroke Volume** (the volume of the ventricle in mL) by the **Heart Rate** (the number of beats per minute). Clearly, the equation says all that in a far shorter and concise way. As an example, if the stroke volume is 90 mL and the heart rate is 70 beats per minute, then

$CO = 90 \text{ mL} \times 70/\text{min} = 6300 \text{ mL/min}$

Please notice that we have been careful to show the units (mL, min) as well as the numbers. They must be there, and indicate to us that Cardiac Output is measured in mL/min.

An algebraic equation also provides us with other information about the items it contains. Consider the following equation, with the units shown in brackets:

area (m²) = length (m) × width (m) $(A = L \times W)$

What happens to the area A if L gets bigger and bigger, but W does not change?

The equation tells us that A and L are related in such a way that the size of A is **directly proportional** to the size of L. That is, if L gets larger, so does A, and if L gets smaller, so does A, and by exactly the same proportion. See Table A1.1.

While this may seem trivial for the case of the area of a square or a rectangle, it is more important when it is applied to something with which you may be less familiar. Think again of the relationship $CO = SV \times HR$. We can now say that, if the heart rate (which is the same as the pulse rate) of the patient declines, the cardiac output will also go down; and that may be useful for you to know if you are responsible for that patient's care.

Now consider the following equation.

$$\text{blood flow} = \frac{\text{blood pressure}}{\text{resistance}}$$

This equation is telling us that blood flow is directly proportional to blood pressure. But in this case, if the resistance **increases**, the flow rate **decreases**. This relationship between blood flow and resistance is referred to as **inversely proportional**. The clue to inverse proportion is given by resistance being in the denominator of the equation.

This information again can be important. If the resistance to blood flow increases, perhaps due to cholesterol deposits in an artery, then the blood flow will be reduced. One way to increase the blood flow in these circumstances is to increase the blood pressure by the same amount as the resistance has increased. The higher than normal blood pressure, which you measure with your sphygmomanometer, is indicating a resistance to flow. This is one reason why blood pressure readings provide valuable clues to internal conditions. See Table A1.2.

Some further examples of algebraic equations are:

$$\text{voltage} = \text{current} \times \text{resistance} \quad V = I \times R$$
$$\text{pressure} = \text{force/area} \quad P = F/A$$
$$\text{density} = \text{mass/volume} \quad D = M/V$$

TABLE A1.1 LENGTH, WIDTH AND AREA

L (m)	W (m)	$A\ (m^2) = L \times W$
2	3	6
4 (2 × 2)	3	12 (6 × 2)
6 (2 × 3)	3	18 (6 × 3)
10 (2 × 5)	3	30 (6 × 5)

TABLE A1.2 BLOOD PRESSURE, RESISTANCE AND FLOW RATE

Blood pressure (BP)	Resistance (R)	Blood flow = BP/R
1	1	1 = 1/1
1	2	0.5 = ½
1	4	0.25 = ¼
1	10	0.1 = 1/10
1	100	0.01 = 1/100

Something a little more complicated:

mean arterial pressure = diastolic pressure + pulse pressure/3   MAP = PP/3

And finally, Poiseuille's Law, which you encountered in Chapter 8 of this text:

$$\text{flow rate} = \frac{(P_2 - P_1) \times \pi r^4}{8vL}$$

PRACTICE PROBLEMS
Percentages
1. Calculate the following:
    (a) 50% of 600
    (b) 15% of 450
    (c) 25% of 1200
    (d) 1% of 15
    (e) 5% of 1000
2. Express as a percentage:
    (a) 1 out of 20
    (b) 5 out of 75
    (c) 12 out of 500
    (d) 78 out of 5000
    (e) 33 out of 1000

Calculations
1. How many litres of dextrose are contained in 3 litres of 10% dextrose?
2. If 2% of a population get heart disease, and the population is 150 000 000 people, how many will get heart disease?

3. How many grams of NaCl are in 500 mL of a 0.9% solution?
4. Find what percentage of a population has a particular blood type if your data show that 5 out of 13 000 have that blood type.
5. A patient is given 85% of an IV bag containing 1200 mL of a 5% dextrose solution. How many mL of solution and how many grams of dextrose did they get?

Ratios

1. If 1 in every 50 people is asthmatic, express this as a ratio.
2. In a class of 76 people, 33 are girls. Express the ratio of boys to girls.
3. Two types of T cells, A and B, occur in the ratio of A:B = 1:4. If there are 550 000 of type A, how many are there of type B?
4. A medication is to be diluted down by 1:5. If we start with 200 mL, how many mL will we end up with?
5. Express 56% as a ratio.

APPENDIX 2
INTRODUCTORY STATISTICS

TYPES OF DATA, DISPLAYING DATA, MEASURES OF CENTRAL TENDENCY AND MEASURES OF DISPERSION

Whenever measurements are made, or data are collected, there is a need to present the findings as clearly as possible to others. Statistics is, in its simplest form, a set of techniques for organising and interpreting measurements. The use of statistics has become so common that some understanding of how it works is essential. Not only is a knowledge of statistics useful for the presentation of your own data, collected in laboratory sessions or research projects, but it is important for making sense of reports and articles written by others.

In this appendix, our concern will be with **descriptive** statistics. These are designed to describe, for your own benefit and others, the specific characteristics of the data. It is concerned with such questions as: how many cases are of a given type; what are the typical values, and what are the extreme values; is there any correlation, or matching, between two sets of data; what is the average value?

TYPES OF DATA

Data which have just been collected, before any organising or interpreting has begun, are referred to as **raw data**. (*Note*: A single piece of collected information is referred to as a **datum**; more than one are called **data**). But even before we start any statistical analysis of the data, it is necessary to determine what type of raw data we have.

First of all, raw data are split into two main types. One type is called **discrete** data, and the other are called **continuous** data. As the names imply, discrete data can only have certain specified values. For example, a person's sex can only be female or male; any birthday can be only one of 365 possibilities. Continuous data, on the other hand, can take on any value within a range. People's weight, for example, can quite possibly take any value from 3 kg upwards. They could be 45.4 kg, or 45.65 kg, or indeed any value we can measure with our scales.

Discrete data

There are two types of discrete data, which we will look at separately—nominal and ordinal data.

Nominal data

These are the data that fall into distinct, separate categories. If we collect information on the sex of a patient, the categories, of course, are male or female. Room number, blood type, and nationality are other examples of nominal data. The categories are not compared with each other in terms of a **value ranking**. That is, it doesn't make any sense to say that a certain blood group, for example, is higher or better or less than another. But they can be compared in terms of which is the more or less common, a **numerical ranking**. Table A2.1 is such a table.

Ordinal data

With ordinal data, we can begin to value rank the measurements, as well as numerically rank them. For example, we can take measurements of patients' reports of pain on a scale like the one that follows.

Scale of severity

1 = mild pain   2 = moderate pain   3 = severe pain

It is then possible to rank the measurements in order of increasing severity of pain. The scale is an arbitrary one with ordinal data; we decided to label mild pain as = 1, when we could have chosen any convenient number. Table A2.2 shows how this may be tabulated.

TABLE A2.1 BLOOD TYPES AND TRANSFUSIONS

Blood type	No. of transfusions per month
A	5
B	17
AB	3
O	27

TABLE A2.2 SEVERITY OF PAIN AND NUMBER OF PATIENTS

Severity of pain	No. of patients
1 (mild)	9
2 (moderate)	60
3 (severe)	13

Continuous data

There are two types of continuous data—interval and ratio data—which we will consider here.

Interval data

Interval data is ranked. That is, it is displayed in a chosen order. Examples would be from smallest to largest, least to greatest, youngest to oldest, or earliest to latest. With interval data, the scale is not quite so arbitrary as it is with ordinal data. To see this, consider an example of measurements of temperature (Table A2.3). While the temperature scale in common uses the Celsius scale, is shared among many millions of people, the choice of the freezing point of water as 0°C was an arbitrary one, picked for convenience. Another example is IQ, where the scale, again, is designed for our convenience, though it is used internationally.

Ratio data

In this last category we find data that can be compared on a scale that does have an absolute, non-arbitrary zero point. This allows us to say, for example, that one measurement is twice as great as another, or one-third as small. This is not possible with interval data, where it makes no sense to say that one person's IQ, for example, is twice that of another person, because we have no measure of

TABLE A2.3 ORAL TEMPERATURE AND NUMBER OF CLIENTS

Oral temperature in °C	No. of patients
35.5	10
36.0	23
36.5	66
37.0	4

what zero IQ refers to. All we can say is that one IQ is higher or lower than another, by a certain amount. With ratio data it makes sense to talk about weight, height and blood pressure, for example, as being zero. And this zero is not done simply for convenience, but represents the complete absence of the quantity we are measuring. See Table A2.4 for an example.

In summary then, raw data can be divided up into two main types—discrete and continuous. In turn, discrete data comes in two forms, nominal and ordinal, while continuous data are found as interval or ratio data.

TABLE A2.4 HEIGHT IN CM AND NUMBER OF PATIENTS

Height in cm	No. of patients
130	5
140	8
150	1
160	23

DISPLAYING DISCRETE DATA

Once the raw data have been collected and classified as to being nominal or ordinal, it is usual to find a way of visually displaying the information. This is done through tables and graphs. Which of these you use depends on your purpose, and which you think is more informative and effective. Since graphs are more pictorial, and can show trends and/or changes in data more clearly than tables, use them if these are your purposes. Tables are often preferred if several items are to be displayed and compared together.

Tables for discrete data

Have a look at the example of a table displaying nominal data (Table A2.5).

TABLE A2.5 FREQUENCY DISTRIBUTION OF GENDER OF PATIENTS UNDERGOING CHEMOTHERAPY OVER 1 YEAR

Gender	f
Males (m)	10
Females (f)	20
	$n = 30$

Notice the following about this table:

1. The table is properly labelled with an appropriate title.
2. f refers to **frequency**; the number of times that value (male or female) occurs in the raw data.
3. n refers to the total number of observations made, and is termed the **sample size**.

A word about the term 'sample size'—as defined above, it refers to the number of observations of the variables that were actually made. The sample does not usually include everyone or everything available for measuring or counting, because often there are too many of them, or we do not have access to them all (for example, no records were kept before a certain year). The total number that theoretically could be measured or counted is referred to as the **population**, and is given the symbol N. So a sample is a group of observations selected from a population.

Graphical displays of discrete data

There are some important features about graphs that are drawn for statistical data that you need to be aware of. As a starting point, we will draw a bar or column graph (Figure A2.1) for the small amount of nominal data given in Table A2.5.

FIGURE A2.1

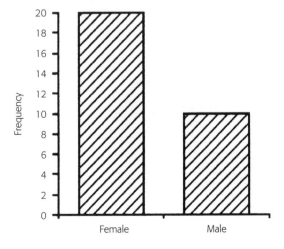

Notice the following features of this graph:

- The frequency f appears on the y-axis.
- The discrete categories appear on the x-axis, and they are the same width.
- The bars or columns *do not touch*; this is because they are discrete.

Another possibility for discrete data is to use a **pie chart** (or pie diagram) to display the data. Here, the frequency for each category is expressed as a percentage of a complete circle. In the example above, $f = 10$ for males and $f = 20$ for females. Expressed as a percentage: $f/n \times 100 = \%$. Here,

$$10/30 \times 100 = 33^1/_3\% \text{ for males}$$
$$20/30 \times 100 = 66^2/_3\% \text{ for females}$$
$$\text{Total} = 100\%$$

On a circle, $100\% = 360°$, so the part of the circle represented by males is

$$33^1/_3 \times 360° = 120°$$

For the females,

$$66^2/_3 \times 360° = 240°$$

This particular example is drawn in Figure A2.2.

FIGURE A2.2

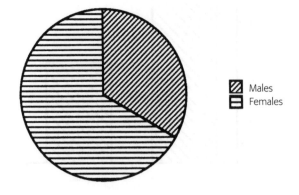

DISPLAYING CONTINUOUS DATA
Tables for continuous data

Here we encounter something quite new. Tables for large numbers of continuous data are often referred to as **grouped frequency distributions**. We do not put each measurement in the table. Rather, we divide the data into equal-sized groups called class intervals, and display the frequency of each class interval in the table. This is perhaps best seen by working through an example.

Imagine that the following raw data were collected—the heights (to the nearest centimetre) of all infants taken into a suburban family health centre for the month of July:

> 75, 67, 76, 71, 73, 86, 72, 77, 80, 75, 80, 96, 93, 75, 73, 83, 81, 82, 73,
> 92, 81, 87, 76, 84, 78, 79, 99, 100, 88, 77, 71, 76, 75, 83, 66, 79, 95, 85.
> ($n = 38$)

In order to put together a grouped frequency distribution from this raw data, the first step is to put it in order, from highest to lowest, and find the frequency of each height. See Table A2.6.

Next, find the **range** of the heights; that is, the difference between the tallest and the shortest. In this case, that is $100 - 66 = 34$. We need the range to help us decide how big our groups, or class intervals, are going to be.

The number of class intervals is up to us to decide, though unwritten rules say it should never be greater than nine. More than nine intervals becomes too difficult to interpret; we might as well have used the raw data. If too few class intervals are used, on the other hand, too many measurements are grouped into

TABLE A2.6 HEIGHT/FREQUENCY TABLE FOR INFANT HEIGHTS FOR JULY

Height	f	Height	f	Height	f	Height	f
100	1	91	0	82	1	73	3
99	1	90	0	81	2	72	1
98	0	89	0	80	2	71	2
97	0	88	1	79	2	70	0
96	1	87	1	78	1	69	0
95	1	86	1	77	2	68	0
94	0	85	1	76	3	67	1
93	1	84	1	75	4	66	1
92	1	83	2	74	0		

each one, and we lose some important details. In our example, suppose we choose seven class intervals.

How many heights should be included in each of our seven class intervals? Since the range is 34, and we have chosen seven class intervals, then each one should hold 34/7 which is 4.86, or about five heights. So the first of our intervals would contain the heights 100, 99, 98, 97, 96 grouped together. The frequencies for each of theses heights are added together to get the frequency of the class interval. We now can make our final table (Table A2.7).

It is easier to interpret the data by looking at the grouped frequency distribution table than it is by inspecting the raw data. The only problem is that some of the precision and detail of the individual measurements has been lost.

Graphing continuous data

There are two common ways of graphing continuous data, and both assume that the measurements have been arranged in a group frequency table like Table A2.7.

Histograms

A histogram is much like a bar or column graph, except that the bars are drawn so that they do touch each other. This is done to show that the measurements are continuous; there are no gaps. The example in Table A2.3 has been drawn using the grouped frequency data of children's heights.

TABLE A2.7 GROUPED FREQUENCY DISTRIBUTION OF CHILDREN'S HEIGHTS

Class interval	f
66–70	2
71–75	10
76–80	10
81–85	7
86–90	3
91–95	3
96–100	3
	$n = 38$

FIGURE A2.3

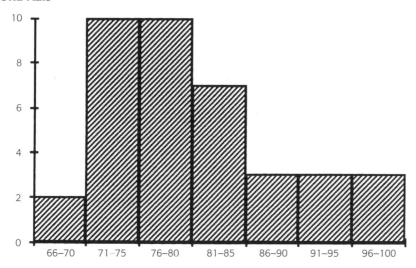

Note the following points:

- The frequency of the class interval is on the *y*-axis.
- The class intervals are shown on the *x*-axis.
- The midpoint of each class interval is at the centre of each bar. (For example, the midpoint of the 66–70 class interval is 68.)
- The endpoints of each class interval lie halfway between the end of one class interval and the beginning of the next. (For example, the real limits of the bar for the 76–80 class interval are 75.5 and 80.5.)

Frequency polygons

The frequency polygon replaces the bars of a histogram with a single line that joins the midpoints of each class interval. Figure A2.4 is an example using the same children's height data as before.

Using frequency polygons, it is often easy to tell if the measurements show any special features. For example, suppose we measure the heights of all 5-year-old children at school in South Australia. We would expect that most children of that age would be roughly the same height, with only relatively few children either a lot taller or a lot shorter. A frequency polygon for such a population might look like the top graph labelled A in Figure A2.5.

Graph A is referred to as a **normal distribution**, and is described as a

FIGURE A2.4

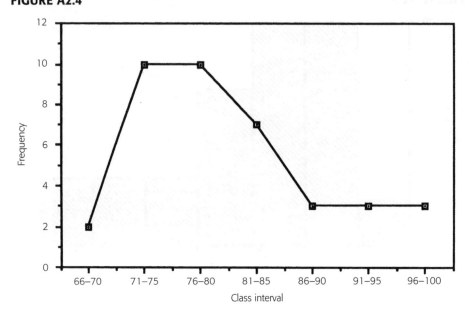

bell-shaped curve. Most of the measurements fall in the middle, somewhere around the average.

If the measurements are not distributed equally around the middle, then the frequency polygon becomes **skewed**. It can take on shapes like the ones labelled B and C in Figure A2.5.

Graph B is called a **negatively skewed polygon**, with most of the measurements being high ones. If these were test scores, for example, it might be telling the teacher that most of the students were very knowledgeable on that topic, or that the test was too easy. With heights, it might be saying that there has been an influx of people who are taller than the average (perhaps due to altered immigration patterns).

Graph C is called a **positively skewed polygon**, with most of the measurements being low ones.

INTERPRETING THE DATA

Once the measurements have been made, and then organised appropriately, it is time to get as much useful information out of them as possible. Here we will

FIGURE A2.5

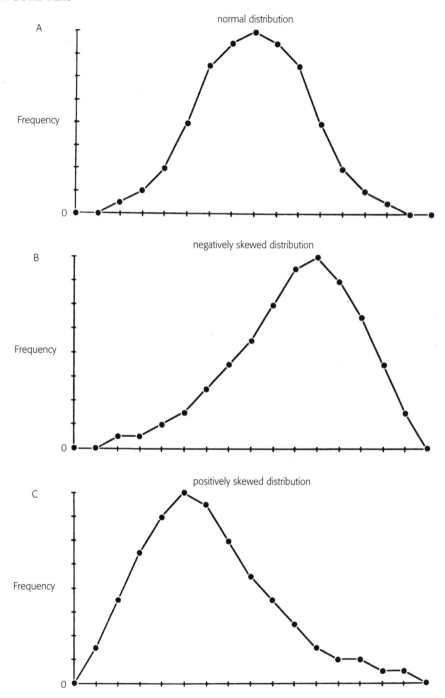

look at two common measures for interpreting data. One is called the **measure of central tendency**. It is designed to help us determine which are the most typical or average measurements. The other is the **measure of dispersion**, which indicates to us the extent to which our measurements are spread out.

Measures of central tendency

You may have heard of these under more familiar names—mode, median and mean. The three all have the same purpose, to help us find average or typical values of our data.

Mode

The mode is simply the most commonly occurring measurement. It can be used with discrete data (for example, the most commonly used room for post-operative recovery). It is the category with the highest frequency in a table, or the tallest bar on a graph, or the largest segment in a pie diagram.

Median

The median is used with ordinal, interval or ratio data. It is that measurement which divides the range of measurements in half. A simple example would be as follows. Suppose we made five measurements of the number of minutes a young child spends watching television at a certain time:

> raw data (time in minutes): 9, 5, 28, 10, 8
> ordered data (in minutes): 5, 8, 9, 10, 28

Here, the median measurement is 9 minutes, because there are two measurements below it (5 and 8) and two above it (10 and 28). Because there were an odd number of measurements, our median was easy to determine.

If the number of measurements is even, then the median must lie somewhere between two of them. In the following example there are six measurements:

> ordered data (in minutes): 6, 17, 19, 20, 1, 27

The method we use is to take the number of measurements, add 1, then divide the total by 2. Since we have six measurements, that gives us $6 + 1 = 7$, and $7/2 = 3.5$

We now count 3.5 from the smaller end of the ordered data. As you can see, that will take us up to 19.5, which is then the median measure. Notice that 19.5 is *not* one of our actual measurements. This makes no difference; it is still the score which divides the measurements into two equal parts.

Mean

The mean is best known as the average of the measurements. That is, we add up all the measurements and divide by the total number of them. The symbol used by statisticians is $\bar{x}$. As a quick example, consider five measurements: 2, 3, 5, 8, 9.

$$\bar{x} = (2 + 3 + 5 + 8 + 9)/5 = 27/5, \text{ and } 27/5 = 5.4$$

Which one of these measures of central tendency we use depends on several factors. First of all, it is obviously nonsense to talk about the mean of discrete data. After all, what could it mean to ask for the average room number, or average sex of a group of patients?

Consider two sets of measurements as shown in Table A2.8.

Both of these samples have the same mode (height = 50), yet they are clearly very different groups of people. Here, the mode does not help us to distinguish clearly between them. For this reason, the mean is often the best measure to use for interval or ratio data, because it uses all the measurements.

However, that rule breaks down when the distribution of the data is highly skewed. Another small example can be used to show why the median is often preferred over the mean when the distribution of a sample or population contains measurements which are quite different from the others.

TABLE A2.8 HEIGHT/FREQUENCY DISTRIBUTION FOR TWO POPULATIONS

Sample A		Sample B	
Height	f	Height	f
50	18	50	10
60	3	60	9
70	3	70	8
80	9	80	6
	$n = 33$		$n = 33$

Sample A: ordered measurements; 2, 2, 2, 5, 7, 8, 9
Mode = 2, median = 5, mean = 5

Sample B: ordered measurements; 2, 2, 2, 5, 7, 8, 44
Mode = 2, median = 5, mean = 10

The trouble with sample B is that unusual measurement of 44; it skews the distribution so much that the mean value for sample B (10) is larger than six of the seven measurements. In short, the mean is too sensitive to the unusual value, and is of less use than the median value.

As a general set of rules, then, we have the following:

☐ The mean cannot be used for discrete data.
☐ The mean is best for continuous data that has a normal distribution.
☐ The median is preferred for continuous data that is highly skewed.
☐ The mode is preferred for discrete data.

Measures of dispersion

It is often important to know how spread out, or dispersed, a group of measurements are. The measures of central tendency cannot tell us anything about this. For example, suppose we have one group of three patients, with weights of 50, 51 and 52 kg respectively. Their mean weight is 51 kg, obviously. But imagine a second group of three, with weights 30, 51 and 72 kg respectively. Again, the mean weight is 51 kg, but clearly this group is quite distinct from the first group in ways that may be important to take into account.

Range

One way to measure the spread of the collected measurements has been briefly mentioned before: the range. We simply subtract the lowest measurement from the highest. For our simple example above, the range of the first group of three patients would be 52 − 50 = 2 kg, which is quite small. For the second group, it would be 72 − 30 = 42 kg, much larger.

The main trouble with the range as a measure is that it depends very strongly on the two end measurements. If one or both of these are rather extreme or unusual, then the range will be misleading. Imagine we had a spread of heights of schoolchildren that showed the distribution 100, 105, 106, 107, 107, 210 cm respectively. The range is 2120 − 100 = 110 cm. This could give a false impression of the differences in size between these six children, five of whom vary by only 7 cm from each other.

Average deviation and standard deviation

Another way to look at the spread of measurements is to ask: how far away, on average, are the measurements from the mean? Tightly grouped measurements will all be found close to the mean. Widely dispersed measurements will be found, on average, further from the mean value. Average deviation tries to answer this question for us. First we find out how far each measurement is from the mean value, then we average those distances. For example, consider the following test scores:

57, 58, 59, 60, 61, 62, 63.

The mean is 60.
Distances from mean:

57 − 60 = −3
58 − 60 = −2
59 − 60 = −1
60 − 60 = 0
61 − 60 = 1
62 − 60 = 2
63 − 60 = 3

If we add up these distances, we get −3 + −2 + −1 +...... = 0
The average deviation from the mean is then 0/7 = 0.

This is the usual result when we have a perfectly symmetrical distribution, and it happens because the negative distances just cancel out the positive distances. Unfortunately, this is not at all useful to us in analysing our data. In order to overcome this problem, it is necessary to first square each of the distances (that is, multiply them by themselves), thus making them all positive numbers. We find the average of these squared numbers, then take the square root of the average. The result is known as a **standard deviation**, symbolised as σ. For the same numbers used in the previous example, we now have:

$$\sigma^2 = ((-3)^2 + (-2)^2 + (-1)^2 + 0^2 + 1^2 + 2^2 + 3^2)/7$$
$$= (9 + 4 + 1 + 0 + 1 + 4 + 9)/7$$
$$= 28/7 = 4$$

The average of the squares of the deviations from the mean is then 28/7 = 4 and the standard deviation σ is $\sqrt{4}$ = 2. This means that, on average, any measurement is two units away from the mean value.

What does the standard deviation really tell us? It can be used to show what percentage of measurements lie within 1 or more standard deviation units from the mean. Tables of these percentages are found in most statistics textbooks. In simple terms, these tables make it possible to say that, for a normal distribution of data, approximately 68% of all the measurements will fall within 1 standard deviation from the mean, and about 95% will fall within 2 standard deviations from the mean (see Figure A2.6).

Clearly, a very small minority of measurements will be found more than 2 standard deviations from the mean.

Measures such as the standard deviation are not only useful for analysing your data. They can be used to help interpret research findings that area given in journals or newspapers or textbooks. If a researcher is making claims about a treatment, for example, it may be important to know if they are discussing individuals or groups who are widely dispersed in some way. Suppose you read about a new therapy that claims to decrease the time for certain types of wounds to heal. Imagine that the mean time for wound healing using current treatments is 25 days, with a standard deviation of 3 days. The new therapy is tried, and the mean healing time is 16 days.

Now, 16 days is a distance of 3 standard deviations from the mean for the current treatment, since 16 = 25 − (3 × 3). Very few patients (less than 5%) would

FIGURE A2.6

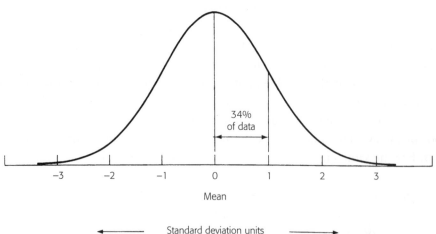

have been expected to show healing in this time using the current treatment. This is therefore good evidence that the new therapy is worth considering.

In order to pull the information in the section together, we will work through a complete example, using wound healing research similar to the above.

EXAMPLE

Raw data: wound healing time (days) under two treatments.

Old treatment (OT): 5, 6, 10, 8, 6, 7, 7, 7, 8, 7, 9, 8, 7, 8, 7, 6, 9, 9, 7, 7, 7, 8, 8, 8.
$N = 24$

New Treatment (NT): 3, 3, 2, 1, 1, 6, 5, 4, 2, 4, 3, 3, 3, 3, 3, 2, 3, 2, 3, 1, 2, 1, 2, 3, 5.
$N = 25$

What type of data is this? It is ratio data.

Now place this data in a table (see Table A2.9). Table A2.10 gives measures of central tendency for both treatments.

TABLE A2.9 NUMBER OF DAYS FOR WOUND HEALING UNDER TWO TREATMENTS

No. of days	OT f	NT f	No. of days	OT f	NT f
1	0	4	6	3	1
2	0	6	7	9	0
3	0	10	8	7	0
4	0	2	9	3	0
5	1	2	10	1	0
			$N =$	24	25

TABLE A2.10 MEASURES OF CENTRAL TENDENCY FOR BOTH TREATMENTS

	OT	NT
Mean (days)	7.46	2.8
Median (days)	7	3
Mode (days)	7	3

Calculation of standard deviation for OT:

$(5 - 7.46)^2 \times 1 = (-2.46)^2 \times 1 = 6.05$
$(6 - 7.46)^2 \times 3 = (-1.46)^2 \times 3 = 6.39$
$(7 - 7.46)^2 \times 9 = (-0.46)^2 \times 9 = 1.90$
$(8 - 7.46)^2 \times 7 = (0.54)^2 \times 7 = 2.04$
$(9 - 7.46)^2 \times 3 = (1.54)^2 \times 3 = 7.11$
$(10 - 7.46)^2 \times 1 = (2.54)^2 \times 1 = 6.45$
Total: 29.94
$\sigma = \sqrt{29.94/24} = \sqrt{1.24} = 1.1$

Calculation of standard deviation for NT:

$(1 - 2.8)^2 \times 4 = (-1.80)^2 \times 4 = 12.96$
$(2 - 2.8)^2 \times 6 = (-0.80)^2 \times 6 = 3.84$
$(3 - 2.8)^2 \times 10 = (0.20)^2 \times 10 = 0.40$
$(4 - 2.8)^2 \times 2 = (1.20)^2 \times 2 = 2.88$
$(5 - 2.8)^2 \times 2 = (2.20)^2 \times 2 = 9.68$
$(6 - 2.8)^2 \times 1 = (3.20)^2 \times 1 = 10.24$
Total: 40
$\sigma = \sqrt{40/25} = \sqrt{1.6} = 1.26$

Figure A2.7 displays both these treatments.

The average of the new treatment, 2.8 days, is just under 3 standard deviations away from the average of the old treatment. According to statistical tables that determine the area of overlap, only about 16% of patients would be expected to recover in 2.8 days using the old treatment.

SUMMARY

This appendix has examined some of the ways descriptive statistics can be used both to present and interpret data. We have classified measurements by the *type* of data: discrete (nominal and ordinal) or continuous (interval and ratio). From there, we explored ways of presenting that data in tables and graphical formats. Finally, we looked at two main ways of interpreting such data: measures of *central tendency* and measures of *dispersal*. With these simple skills, it should be possible to meaningfully present and describe the results of laboratory practicals, measurements collected in community or hospital settings, or the data given by others in your readings.

FIGURE A2.7

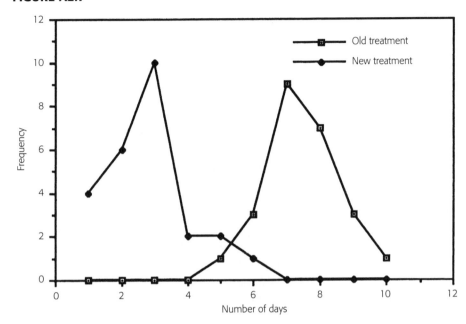

GLOSSARY

Accommodation The changing in shape of the lens of the eye to allow it to focus the light from objects at varying distances from the eye.

Acid Molecules that release an H^+ ion when they dissociate.

Acid/base buffering The chemical homeostatic control of the levels of H^+ and OH^- in the body.

Acid/base neutralisation The chemical reaction between an acid and a base, in which they form products (usually a salt and water) which are neither acidic nor basic.

Acidosis The condition of excess H^+ ions in the body.

Action potential The stimulus that makes a neuron fire or depolarise.

Active transport The movement of solute particles across a cell membrane, using the cell's energy. It may move particles from low to high concentrations.

Alkalosis The condition of excess OH^- ions in the body, or lowered levels of H^+ ions.

Amplitude For a water wave, its height above level surface; for a sound wave, the size of an area of compression or rarefaction. In the case of light, amplitude is related to brightness; for sound, it is related to loudness.

Astigmatism A condition causing blurred vision, caused by unevenness of the cornea and/or lens. Refraction does not occur uniformly.

Atom The smallest unit of an element. Since there are 92 naturally occurring elements, there are 92 naturally occurring types of atoms.

ATP and ADP Acronyms for adenosine triphosphate (the source of energy used by all cells) and adenosine diphosphate.

Average deviation A measure of dispersion, it indicates how far on average any given value is from the mean; see also standard deviation, Appendix 2.

Base (or alkali) A molecule that both accepts an H^+ ion during a chemical reaction and releases an OH^- ion (hydroxide ion) when it dissociates.

Becquerel (Bq) The unit of radiation measuring the number of decays per second; 1 Bq = 1 decay/second.

Biological half-life The time taken for one-half of a given amount of a radioactive substance to be eliminated from the body by normal excretion.

Buffer systems See Chapter 7. Examples:

Bicarbonate buffer system Chemical homeostatic control of H^+ ion concentration using the dissociation of carbonic acid, H_2CO_3.

Haemoglobin buffer system Chemical homeostatic control of H^+ ion concentration using haemoglobin to transport H^+ to the lungs.

Phosphate buffer system In cells, using the dissociation of $H_2PO_4^-$ ions.

Protein buffer system In cells, where amino acids accept or release H^+ ions.

Boyle's Law The pressure of a gas is inversely proportional to the volume it is confined in, providing the temperature remains unchanged.

Catabolism and anabolism Two types of metabolic reactions. Catabolic reactions break large molecules into smaller ones; anabolic reactions build larger molecules out of smaller ones.

Cardiac output The amount of blood being circulated around the body; for an adult, about 5 litres/min. Cardiac output = pulse rate × stroke volume

Cell The body's smallest independent unit of life. Cells have complex structures that allow them to carry out their diverse functions.

Cellular respiration The process of converting the energy in carbohydrate or lipid molecules into ATP. There are four steps in this process.

Charles' Law The volume of a gas is directly proportional to the temperature, in Kelvin, providing the pressure remains unchanged.

Chemical reaction The term used for the joining together of atoms, ions or molecules and for the splitting apart of molecules into either atoms, ions or smaller molecules.

Codon A combination of three nucleotide bases; a codon 'codes' for one specific amino acid.

Coenzymes and cofactors Two of the factors that assist enzymes in their function. Coenzymes are the non-protein part of the enzyme molecule (e.g. vitamin C); cofactors are often metal ions (e.g. Zn^{+2}) which occur at the enzyme binding site.

Colloid A solution in which the solute particles are not dissolved, yet are too small to sink to the bottom.

Compound One or more atoms of different types joined together chemically. Strictly speaking, the atoms are joined in a fixed proportion of one to another. CO_2 is a molecule of a compound; H_2 is a molecule of an element.

Conductor A material (solid, liquid or gas) that allows the free movement of charged particles (current electricity) through it. For example, electrodes, which can be used to detect the presence of charges, or conduct it to any particular point.

Continuous data Data which can have any value in a range or spectrum, such as height or mass.

Covalent bond Type of chemical bonding in which atoms share orbiting electrons. This sharing is done so that the atoms will have enough electrons to fill the outermost energy level of each atom.

Current (*I*) The number of charged particles that move past a given point in a second, measured in amperes. In relation to voltage (which supplies the energy for the movement) and resistance (which tends to prevent the movement), $I = V/R$.

Data Information collected by various means, such as measurements or interviews. Data come in several forms; see also raw data, discrete data and continuous data.

Deamination The removal of amine groups (those containing nitrogen) from amino acids. It is a step in the breakdown of protein for energy purposes.

Decibel One-tenth of a bel, the unit used to compare the power of two sounds. It is also used to indicate loudness.

Dehydration reaction The opposite of an hydrolysis reaction; in this case, one of the products is water, the elements of which have been removed from one or more of the reactants.

Depolarisation The response of a neurone to an action potential, during which it changes its electric potential from −70 mV to +30 mV in about 1 ms.

Derived units Units that are combinations of two or more of the seven basic SI units.

Dialysis The use of diffusion and/or osmosis to remove unwanted solutes/solvents from a solution.

Diastolic pressure Lowest blood pressure, during the relaxation of the heart muscle between contractions.

Diffusion The movement of particles from an area of higher concentration to an area of lower concentration. It may or may not occur through a permeable membrane. Diffusion is passive (requires no outside energy) and results from the random motion of the particles.

Diffusion coefficient A measure of the ease with which a particular molecule will diffuse through a particular membrane; the higher the coefficient, the greater the amount of diffusion.

Direct radiation damage Damage done by the physical impact of radiation on a molecule.

Discrete data Data which can have only certain specified values, such as blood type.

Dissociation The breakdown of a solute into ions or smaller molecular units.

DNA The abbreviation for deoxyribonucleic acid. It is the unit of heredity, containing genes that code for all the proteins essential for life.

Dominant gene The gene in a heterozygous pair whose character is expressed in the organism; i.e. brown eyes genes are dominant over blue eyes genes, so the individual has brown eyes.

Earthing, or grounding Providing a pathway to the earth for charged particles; providing a safety control of static and current electricity.

Effector Structures in the body which respond to signals from a sensor or an integrating centre like the brain, by altering their normal function (e.g. muscles shivering).

Electrocardiogram The record of electrical changes in heart activity that are detected by an electrocardiograph.

Electrocardiograph (ECG) A machine that detects the electrical activity of the heart and displays it on an oscilloscope screen and as a graph on paper.

Electroconvulsive therapy (ECT) The application of low levels of electric current to the brain in an attempt to treat depression.

Electroencephalogram The record of electrical changes in the brain that are detected by an electroencephalograph.

Electroencephalograph (EEG) A machine for detecting and displaying the electrical activity of the brain. Typical voltages recorded range from 0 to 300 mV.

Electrolytes Ions in solution capable of conducting an electric current through that solution.

Electron The particle that circles the nucleus. Much smaller than either the proton or neutron, it has a –1 charge.

Electron transport chain A step in cellular respiration, in which 34 ATP molecules are produced, using hydrogen from the Krebs cycle and oxygen from breathing.

Electrostatic force The force of attraction and/or repulsion between electric charges. Opposite charges attract (+ to –) and like charges repel. It holds the electrons to the nucleus and holds ions of different charges together in ionic bonds.

Element A substance that cannot be broken down into any simpler substance. There are 92 naturally occurring elements, about 16 of which are crucial for healthy body function.

Energy Often defined as the ability to do work and responsible for causing change. It is measured in joules. There are many types, but the two basic forms are kinetic energy and potential energy.

Enzymes Protein molecules that increase the speed of a particular chemical reaction. Their effect is linked to their specific shape.

Equilibrium A balance of forces, so that the object does not move, or moves with a steady speed. A client in bed is in equilibrium, balanced by gravity (downwards) and the support of the bed (upwards).

Facilitated diffusion Diffusion in which the solute particles are assisted through the cell membrane by carrier molecules designed for that purpose.

False positive result An error in clinical measurement that implies certain chemicals are present in the sample when they in fact are not, or are there for reasons that have not been taken into account (e.g. medication).

Fibrillation The random depolarisation of heart muscle tissue, due to loss of control by the sinoatrial node. It is sometimes treated with an electric current from a defibrillator.

Fission The splitting of the nucleus of a radioactive isotope into two or more fragments, along with the release of a great deal of energy.

Fluoroscopy A diagnostic technique using a continuous stream of X-rays to produce a moving picture of structures in the body.

Force In simple terms, a push or a pull done in an effort to get an object to change its motion. Technically, it's the product of the mass of the object (in kg) and its acceleration (in m.s^{-2}); force = mass × acceleration. It is measured in newtons (N).

Frequency The number of waves per second, measured in hertz (Hz). The product of frequency and wavelength is the speed of the wave.

Frequency polygons A graph of continuous data that shows a single line joining the midpoint of each class interval. They are classified as showing either a normal distribution, or skewed distribution; see Appendix 2.

Functional group An atom or group of atoms that has a strong influence on the chemical behaviour of an organic compound; examples are –OH in alcohols, and NH$_2$ in amines.

Gene A string of codons that code for a particular protein.

General Gas Law Combines Boyle's Law and Charles' Law into one expression. Mathematically, for a given sample of a gas $P_1V_1/T_1 = P_2V_2/T_2$. Here, P_1 means the pressure of the gas before any changes were made and P_2 is the pressure after changes occurred; likewise for V and T.

Glycogenesis The process of converting glucose to glycogen; it occurs in the liver.

Glycogenolysis The process of converting glycogen back into glucose; it occurs in the liver.

Glycolysis One step in cellular respiration, in which glucose is converted to pyruvic acid and two ATP molecules are produced.

Gray (Gy) The unit of radiation that measures how much energy is given to the material absorbing it.

Henry's Law The amount of any gas that will dissolve in a liquid is proportional to its partial pressure and its solubility coefficient.

Heterozygous The genes in the pair, one from each parent, are different; e.g. one codes for a blue eyes protein, one for a brown eyes protein.

Histogram A bar or column graph of continuous data, where the columns contact each other.

Homeostasis The term used to describe the complex and numerous activities carried out by the body to maintain a relatively constant internal environment (e.g. relatively constant temperature).

Homozygous The condition where both genes for a given protein, one from each parent, are the same; e.g. both code for a protein responsible for blue eyes.

Hydrogen bond A weak attraction between a hydrogen and a nearby oxygen atom.

Hydrolysis reaction One in which water molecules take part; when added to a reactant, water breaks it down into smaller products.

Hypertonic A solution in which the concentration of a given solute is greater outside the cell than inside.

Hypotonic A solution in which the concentration of a given solute is less outside the cell than inside.

Infrared thermography A diagnostic technique which forms images of structures using the infrared radiation they emit.

Interval data Continuous data that is ranked in an arbitrary way, like youngest to oldest.

Ion An atom or molecule that has either gained or lost one or more electrons. If it has gained, it is referred to as a negative ion; if lost, it's a positive ion.

Ionic bond Type of chemical bonding between ions, where positive ions are held together with negative ions by the electrostatic force of attraction.

Indirect radiation damage Damage done to a cell by the ionisation reactions that occur due to the passage of ionising radiation through that cell.

Insulator The opposite of a conductor; it prevents or slows down the flow of current electricity.

Isotonic A solution in which the concentration of a given solute is the same on either side of the semi-permeable membrane surrounding a cell.

Isotope A variety of atomic nucleus. The isotopes of an element all have the same number of protons, but differ in the number of neutrons present.

Kinetic energy The energy of objects in motion. It is found by $E_k = 1/2 \times \text{mass} \times \text{velocity}^2$.

Krebs cycle A step in cellular respiration. Carbon dioxide and hydrogen are given off and two ATP molecules are produced.

LD$_{50}$ The amount of radiation that will kill 50% of a population within a specified time.

Lever A combination of load, force and fulcrum. Levers allow either greater strength (first and second class), or a greater range of movement (third class).

Macroshock Electric shock involving high voltages and currents, passing through the skin.

Magnetism Caused by the movement of electric charges, it results in the establishment of north and south magnetic poles. Living organisms are generally magnetically neutral, as the two poles always occur together.

Mean A measure of central tendency, better known as the average value.

Measurement A number plus a unit. Measurements are a comparison of one aspect of an object, such as its length, with an agreed upon standard; in this case, the metre.

Median A measure of central tendency, it is the measurement which divides the range of measurements in half.

Metabolic acidosis Loss of bicarbonate ion from the body, or an excess of acids taken into, or produced by, the body.

Metabolic alkalosis An increase in alkali into the body, or loss of acid by the body (e.g. by vomiting).

Metabolism The total of all the chemical reactions that occur in the body. These reactions are also responsible for the heat energy that maintains our body temperature.

Microshock A type of electric shock involving very small voltages and very small currents, which gain entry to the body by avoiding the skin's protective high resistance.

Milliequivalent solutions (mEq/L) Solution concentrations that indicate the number of moles of ion solute times the number of the electric charge on the ion, divided by 1 litre of solvent.

Mixture A grouping together of elements, compounds and/or molecules which are not chemically joined by bonds and are not joined in a fixed proportion, for example air or a mixture of sand and salt.

Mode A measure of central tendency, it is the most commonly occurring measurement.

Molar solutions Solution concentration expressed as moles of solute dissolved in a litre of solvent.

Mole The SI unit of quantity. A mole is equal to the atomic weight (or gram molecular mass) of the element (or molecule) expressed in grams and contains Avogadro's number of atoms, about 6×10^{23}.

Molecule One or more atoms joined together chemically, for example H_2 and CO_2.

Mutation A change in the codon sequence in a DNA molecule. It may alter the amino acid sequence and therefore the proteins made.

Myopia Near or shortsightedness, caused by either an eyeball that is too deep or the inability of the lens to elongate fully (accommodate); images are focused in front of the retina. It is corrected with concave lenses.

Negative feedback loop A common homeostatic control mechanism. A stimulus, or change away from normal in the internal environment (e.g. a decrease in blood temperature), leads to a particular action being taken by the body (e.g. shivering). This action by the effectors makes the detected change decrease (e.g. the blood temperature rises towards normal). This decrease in the change causes a reduction in the body's reaction to the change (e.g. shivering slows down, then stops).

Nominal data Discrete data that fall into distinct, separate categories.

Normal range Those values for any measurement that indicate the expected, usual variation between individuals. Measurements whose values fall outside this normal range are taken to be indicators of an abnormal condition. How wide or narrow the normal range for any measurement should be is not always easy to determine, however. Tables of normal values are published by relevant authorities.

Nuclear radiation The particles, including gamma photons, that are emitted from the nucleus of a radioisotope. There are three types:
alpha rays: the nucleus of a helium atom
beta rays: an electron
gamma rays: high energy photons of light.

Nucleotide The building-block of the DNA molecule, it contains three parts: sugar, phosphate and one nitrogen base.

Nucleus The central region of an atom, consisting of one or more protons (with a +1 charge) and one or more neutrons (with no charge).

Ordinal data Discrete data that can be ranked, such as levels of pain.

Organ A combination of two or more tissues into one structure, carrying out a single function (e.g. the heart).

Organic molecule One containing carbon and which usually, but not necessarily, has been made by a living organism, for example carbohydrates, lipids and proteins.

Osmoles and osmolarity Osmolarity is measured in osmoles. It is the number of moles of solute times the number of particles the solute dissociates into upon dissolving. Osmoles are so named because the number of solute particles is critical for the process of osmosis.

Osmosis The movement of solvent molecules (usually water) from areas of higher solvent concentration to lower solvent concentration. The movement is passive and occurs across a membrane permeable to the solvent.

Osmotic pressure The solvent pressure that builds up as a result of the accumulation of solvent (say, in a cell) caused by osmosis. Osmotic pressure can be great enough to prevent further osmosis from occurring.

Oxidation/reduction reactions Two important types of chemical reactions that occur in the body. Oxidation reactions are those in which (1) there is a loss of electrons and/or (2) there is a loss of hydrogen or a gain in oxygen. Reduction reactions are those in which (1) a gain of electrons and/or (2) a gain of hydrogen or loss of oxygen. These reactions are particularly crucial in cellular respiration.

Oxidative phosphorylation The chemical process by which a phosphate bond is added to an ADP molecule to make a molecule of ATP. The energy for this process is stored in a molecule of glucose.

Partial pressure The pressure exerted by a specific gas, as a contribution to the total pressure of all the gases present. For a gas X, it is written p_x pascals.

Parts per million (ppm) and parts per billion (ppb) A measure of concentration, usually used in discussing pollutants. Technically, 1 ppm = 1 mg/kg and 1 ppb = 1 µg/kg.

Pascal's Principle Any change in pressure anywhere in a closed system is transmitted undiminished to all parts of the system.

Per cent solutions The ratio of solute to solvent expressed as a percentage; i.e. 5% dextrose = 5 g dextrose/100 mL solution.

Permeable (or semi-permeable) The property of a membrane that allows the diffusion of certain particles to occur. Permeability refers to the presence of 'pores' in the membrane through which molecules of the right size, shape and electric charge can pass.

Persistence/residence time Indicators of how long a particular toxin or pollutant remains in the environment (or in the body) before being broken down to a harmless chemical. Residence time usually refers only to gases or vapours in the atmosphere.

Physical half-life The time taken for one-half of a given amount of a radioactive substance to decay into either stable or unstable forms.

Plasma or serum The liquid part of blood, containing water and all dissolved solutes. Technically, serum is plasma minus the factors responsible for clotting the blood.

Polar bonds Bonds between pairs of atoms which result in the molecule having one end with a slight negative charge and the other end with a slight positive charge. In water, for example, the oxygen end of the molecule more strongly attracts the shared covalent electrons, giving that end the slight negative charge. Two such water molecules can hold together because of the electrical attraction between them; this is a hydrogen bond.

Pollutant A harmful xenobiotic that is either a naturally occurring inorganic chemical (such as asbestos) or a product of human technology (such as DDT or petrol fumes).

Potential energy The energy of position, usually taken to mean how high from the ground, or how far apart. It's found by E_p = mass × acceleration × position.

Presbyopia and hyperopia Farsightedness, caused either by an eyeball that is too narrow (hyperopia) or the loss of elasticity by the lens (presbyopia); images are focused behind the retina. They are corrected with convex lenses.

Pressure A measure of force per area; P = force/area. It is measured in pascals, though mm Hg is still used in nursing; see Chapter 1.

Radiography The diagnosis and/or treatment of a tumour by radiation. It may be internal (sealed or unsealed) or external.

Radioisotope An isotope that is radioactive; it emits radiation from an unstable nucleus.

Radiosensitivity Indicates the susceptibility to destruction by radiation of a particular type of tissue or tumour.

Ratio data Continuous data that can be compared on a scale with a fixed, non-arbitrary zero point, like mass or height.

Ratio solutions Solution concentration expressed as the ratio of solute to solution; e.g. as 1:100.

Recessive gene The gene in a heterozygous pair whose character is not expressed in the organism. In a homozygous recessive individual, however, that gene is expressed; e.g. an individual with two blue eyes genes will have blue eyes.

Refraction The change in direction of a wave due to a change in speed. For light, this allows an image to be focused on the retina.

Repolarisation The process of returning the neuron to its resting potential after it has depolarised.

Resistance (R) Any opposition to the free movement of electric charges, measured in ohms (Ω).

Resonance A physical property of some materials that allows them to vibrate at the same frequency as a source of sound with which they are in either direct or indirect contact.

Respiratory acidosis Excess H^+ ion concentration due to hypoventilation, causing a build-up of CO_2.

Respiratory alkalosis Low levels of H^+ ion, caused by hyperventilation flushing CO_2 from the body.

Resting potential The state of voltage of a neuron when it's at rest, just before being stimulated, usually about -70 mV.

Reversible reaction One that can take place in both directions; that is, reactants form products at the same time as products reform the reactants. These reactions are indicated by double-headed arrows: $\rightleftarrows$.

RNA The abbreviation for ribonucleic acid. Available in three forms, it translates the DNA code into protein at the ribosomes.

Saturated and unsaturated fatty acids Saturated fatty acids have no double bonds between the carbon atoms in the chain; thus they are saturated with hydrogen atoms. Unsaturated fatty acids have fewer hydrogen atoms, because some of the bonds are in the form of double bonds between carbon atoms.

Sensor A structure in or on the body which responds to a change in its environment by signalling to other organs or systems (e.g. the hypothalamus and temperature, the eye and light).

Sievert (Sv) The unit of radiation that measures the damage done by the absorbed dose.

SI SI stands for System International; its formal (French) title is Système International d'Unités. The seven standards of comparison in this system are mainly based on the behaviour of atoms. It is a decimal system, based on powers of 10, and it is also referred to as a metric system, based on the metre.

Solubility coefficient A measure of how easily a gas dissolves in water at a known temperature; the higher the coefficient, the more gas dissolves.

Solution The combination of a solute (or solutes) dissolved in a solvent. The solutes are the substances (gases, liquids or solids) that are dissolved in the solvent, which is usually water.

Static electricity An accumulation of excess positive or negative charges, which can discharge through a spark; it is particularly dangerous in microshock.

Strong nuclear force One of the four fundamental forces of nature, it holds the protons within the nucleus, opposing their electrostatic repulsion.

Surface tension A property of liquids. The surface appears to be covered with a thin elastic membrane caused by unbalanced cohesive forces among the molecules at the surface. Surface tension pulls water into spherical drops.

Surfactant A substance that can reduce the surface tension of its solvent, or the tension formed between one surface (such as water) and another (such as the walls of an alveolus).

Suspension A solution in which the solute particles are not dissolved and are large enough to settle to the bottom over time.

System A combination of organs carrying out a major task, such as the circulatory system. Organs may be part of more than one system.

Systolic pressure Highest blood pressure, a result of heart contraction.

Tissue A collection of cells of the same type, performing the same function (e.g. muscle tissue).

Tomography A diagnostic technique using X-rays, ultrasound, or radio frequency electromagnetic waves, generated from a moving source; a three-dimensional image can be formed by using computer technology.

Torque The tendency of a force to produce rotation around a pivot point, as in untwisting a jar lid.

Toxin A harmful xenobiotic that is the product of plant or animal metabolism.

Ultrasound Sound waves of frequencies higher than those detectable by the human ear; i.e. over 20 000 Hz.

Ventilation Commonly called breathing; the movement of air into and out of the lungs by inhaling (inspiring) and exhaling (expiring).

Volts (V) Technically, energy/charge. The energy comes from the work required to separate oppositely charged particles (e.g. the work done by neurones using active transport) and the charge is the charge on the ions that are moved. This is also referred to as electric potential, as the energy is in the form of potential energy (e.g. the neurone's negative electric potential of –70 mV).

Wavelength The distance between one part of a wave to the corresponding part on the succeeding wave. For light, an indicator of colour; for sound, an indicator of pitch.

Weak nuclear force This fundamental force is responsible for holding the neutron together. Outside the nucleus of an atom, a neutron spontaneously splits into a proton and an electron.

Work The product of force times distance moved. The work done by a muscle, for example, is the force applied by the muscle times the distance it moves. W = force × distance. It is measured in joules.

BIBLIOGRAPHY

Ackerman, V. and Dunk-Richards, G. (1991) *Microbiology: An Introduction for the Health Sciences*, W.B. Saunders, Sydney.

Astrand, P. and Rodahl, K. (1986) *Textbook of Work Physiology*, 3rd edn, McGraw-Hill, New York.

Australian Academy of Science (1985) *Biological Science: The Web of Life*, 3rd edn, Australian Academy of Science, Canberra.

Brafield, A.E. and Llewellyn, M.J. (1982) *Animal Energetics*, Blackie, New York.

Brown, J.E. (1990) *The Science of Human Nutrition*, Harcourt Brace Jovanovich, San Diego.

Brown, W.H. (1975) *Introduction to Organic Chemistry*, Willard Grant Press, Boston.

Clark, W.G. et al. (1988) *Goth's Medical Pharmacology*, 12th edn, Mosby, St. Louis.

Commonwealth Department of Health (1980) *Hospital Diet Manual*, Australian Government Publishing Service, Canberra.

Corbett, J.V. (1992) *Laboratory Tests and Diagnostic Procedures with Nursing Diagnoses*, 3rd edn, Appleton & Lange, New York, USA.

Curtis, H. and Barnes, N.S. (1985) *Invitation to Biology*, 4th edn, Worth Publishers, New York.

Donovan, E.W. (1985) *Essentials of Pathophysiology*, Macmillan, New York.

Emery, A.E.H. and Mueller, R.F. (1988) *Elements of Medical Genetics*, 7th edn, Churchill Livingstone, Melbourne.

Gatsford, J.D. (1987) *Nursing Calculations*, 2nd edn, Churchill Livingstone, Melbourne.

Giancoli, D.C. (1991) *Physics*, 3rd edn, Prentice Hall, New Jersey.

Giancoli, D.C. (1998) *Physics*, 5th edn, Prentice Hall, New Jersey.

Gordon, J.E. (1976) *The New Science of Strong Materials*, 2nd edn, Penguin, London.

Guyton, A.C. (1986) *Textbook of Medical Physiology*, W.B. Saunders, Philadelphia.

Hill, D.W. and Dolon, A.M. (1976) *Intensive Care Instrumentation*, Academic Press, London.

Hobbie, R.K. (1988) *Intermediate Physics for Medicine and Biology*, 2nd edn, John Wiley & Sons, New York.

Huber, G.L. (1978) *Arterial Blood Gas and Acid-Base Physiology*, Upjohn, USA.

Manahan, S.E. (1989) *Toxicological Chemistry*, Lewis Publishers, Michigan.

Martini, F. (1992) *Fundamentals of Anatomy and Physiology*, 2nd edn, Prentice Hall, New Jersey.

Martini, F. (2001) *Fundamentals of Anatomy and Physiology*, 5th edn, Prentice Hall, New Jersey.

Maxon, L.R. and Daugherty, C.H. (1989) *Genetics: A Human Perspective*, 2nd edn, Wm. C. Brown, Iowa.

McCance, K.L. and Huether, S.E. (1980) *Pathophysiology*, Mosby, St Louis.

McMurry, J. (1989) *Essentials of General, Organic and Biological Chemistry*, Prentice Hall, New Jersey.

McMurry, J. and Castellion, M.E. (1992) *Fundamentals of General, Organic and Biological Chemistry*, Prentice Hall, New Jersey.

Pagana, K.D. and Pagana, T.J. (1990) *Diagnostic Testing and Nursing Implications: A Case Study Approach*, 3rd edn, Mosby, St. Louis.

Porth, C.M. (1986) *Pathophysiology*, 2nd edn, J.B. Lippincott, Philadelphia.

Potter, P.A. and Perry, A.G. (1987) *Basic Nursing: Theory and Practice*, Mosby, St. Louis.

Rossotti, H. (1983) *Colour*, Penguin, London.

Sackheim, G.I. and Lehman, D.D. (1981) *Chemistry for the Health Sciences*, 5th edn, Macmillan, New York.

Stollberg, R. and Hill, F.F. (1975) *Physics: Fundamentals and Frontiers*, Houghton Mifflin, Boston.

Taylor, R. (1979) *Noise*, 3rd edn, Pelican, London.

Thompson, M.W. et al. (1991) *Genetics in Medicine*, 5th edn, W.B. Saunders, Philadelphia.

Tillery, B.W. (1991) *Physical Science*, Wm. C. Brown, Iowa.

Tillery B.W. (1992) *Introduction to Physics and Chemistry: Foundations of Physical Science*, Wm C. Brown, Iowa.

Timberlake, K.C. (1988) *Chemistry*, 4th edn, Harper and Row, New York.

Timberlake, K.C. (1992) *Chemistry*, 5th edn, HarperCollins, New York.

Timberlake, K.C. (1999) *Chemistry*, 7th edn, HarperCollins, New York.

Van de Graaff, K.M. and Fox, S.I. (1989) *Concepts of Human Anatomy and Physiology*, 2nd edn, Wm. C. Brown, Iowa.

INDEX

Headings in **bold** indicate terms which are also defined in the glossary; page numbers in *italics* indicate figures

ratio data 313–14, 322
ratio solutions 172
ratios 305–6
raw data 311
reactants 68
recessive genes 261, *262*
reciprocals 8
red blood cells 21, *182–3*, 182–4, 197
reduction reactions 112
refraction 233–9, *237–8*
relative atomic mass 173
rem 289
renal dialysis, case study 154, 158–9, 163–4, 171
renal scans 300
repolarisation 214–15, 219
residence time 275
residual volume, of gases 132–3
resistance 211–12
resonance 247–8
respiration 110, 125–32, *126–7*, *131*, 142–4, *144–5*, 199–200
see also cellular respiration
respiratory acidosis 200
respiratory alkalosis 200–1
respiratory system 19
response, defined 26
resting potential 212–13
retina *236*
reversible reactions 69
rhodopsin 235
ribonucleic acid 92–3, *257*, 262–3, *265–7*, 266–7
ribosomal RNA 263
ribosomes *17*
RNA 92–3, *257*, 262–3, *265–7*, 266–7
rods 235

S
SA node cells 208–11, 219, 221, 222
salt 175
sample size 315
saturated fatty acids 84, *84*
scoliosis 268
sealed internal radiography 297–8
secondary structure, of protein 89, 91
second-class levers *45*, 46–7
secretory vesicles *17*
seminoma of the testis 299
semi-permeable membranes 140–1, *157*, 161, 162
sensors 21–3, *22*
serum 181–4

shortsightedness 239, *240*
SI system 4, 6
sickle cell anaemia 268
sievert 289
sinoatrial node cells 208–11, 219, 221, 222
skewed distributions 320, *321*
skin pigmentation, effect of radiotherapy on 299
smog 147, 275–7
snake venom 274
sodium chloride *156*, 175
sodium ions 64, 65, *157*
solubility coefficients 134
solubility, relation of to pressure 134–6, *135*
solutes 138, 155–6, *156*, 161, 164, 185–6, *186*
solutions
blood as a solution 138–9, 141, 181–7, *182–3*
concentrations of 169–74
hypertonic solutions *183*, 183–4
hypotonic solutions *183*, 183–4
isotonic solutions *182*, 182–3
milliequivalent solutions 174–5
molar solutions 172–4, *173*
osmole and milliosmole solutions 175
per cent solutions 172
ratio solutions 172
solvents 138, 155–6, *156*, 161, 164
sound
medical applications of 248–50
noise pollution 250–1
production of *244*
resonance 247–8
sound waves *244*, 244–7
speed of 246
sphygmomanometer 308
spider venom 274–5
spiral fractures *50*
standard deviation 325–8, *326*, *329*
starch 78
static electricity 210–11
statistics 311–29
stearic acid 82–4, *83*
steroids 81, 86
stimulus, defined 26
stress 26, 251
strong nuclear force 63, 285
structure, of body parts 15
substrates *103*, *105*, 106
sucrase 90
sucrose 77, 78, *79*
sugar 258

THE PERIODIC TABLE

Legend:
- 1 — Atomic Number
- H — Chemical Symbol
- Hydrogen — Element
- 1.0079 — Atomic Weight

1	2	3	4	5	6	7	8	9	10	11	12	13	14	15	16	17	18
1 **H** Hydrogen 1.0079																	2 **He** Helium 4.00260
3 **Li** Lithium 6.941	4 **Be** Beryllium 9.01218											5 **B** Boron 10.81	6 **C** Carbon 12.011	7 **N** Nitrogen 14.0067	8 **O** Oxygen 15.9994	9 **F** Fluorine 18.99840	10 **Ne** Neon 20.179
11 **Na** Sodium 22.98977	12 **Mg** Magnesium 24.305											13 **Al** Aluminium 26.98154	14 **Si** Silicon 26.086	15 **P** Phosphorous 30.97376	16 **S** Sulphur 32.06	17 **Cl** Chlorine 35.453	18 **Ar** Argon 39.948
19 **K** Potassium 39.098	20 **Ca** Calcium 40.08	21 **Sc** Scandium 44.9559	22 **Ti** Titanium 47.90	23 **V** Vanadium 50.9414	24 **Cr** Chromium 51.996	25 **Mn** Manganese 54.9380	26 **Fe** Iron 55.847	27 **Co** Cobalt 58.9332	28 **Ni** Nickel 58.71	29 **Cu** Copper 63.546	30 **Zn** Zinc 65.38	31 **Ga** Gallium 69.72	32 **Ge** Germanium 72.59	33 **As** Arsenic 74.9216	34 **Se** Selenium 78.96	35 **Br** Bromine 79.904	36 **Kr** Krypton 83.80
37 **Rb** Rubidium 85.4678	38 **Sr** Strontium 87.62	39 **Y** Yttrium 88.9059	40 **Zr** Zirconium 91.22	41 **Nb** Niobium 92.9064	42 **Mo** Molybdenum 95.94	43 **Tc** Technium 98.9062[b]	44 **Ru** Ruthenium 101.07	45 **Rh** Rhodium 102.9055	46 **Pd** Palladium 106.4	47 **Ag** Silver 107.868	48 **Cd** Cadmium 112.40	49 **In** Indium 114.82	50 **Sn** Tin 118.69	51 **Sb** Antimony 121.75	52 **Te** Tellurium 127.60	53 **I** Iodine 126.9045	54 **Xe** Xenon 131.30
55 **Cs** Cesium 132.9054	56 **Ba** Barium 137.34	57* **La** Lanthanum 138.9055	72 **Hf** Hafnium 178.49	73 **Ta** Tantalum 180.9479	74 **W** Tungsten 183.85	75 **Re** Rhenium 186.2	76 **Os** Osmium 190.2	77 **Ir** Iridium 192.22	78 **Pt** Platinum 195.09	79 **Au** Gold 196.9665	80 **Hg** Mercury 200.59	81 **Tl** Thallium 204.37	82 **Pb** Lead 207.2	83 **Bi** Bismuth 208.9804	84 **Po** Polonium (210)[a]	85 **At** Astatine (210)[a]	86 **Rn** Radon (222)[a]
87 **Fr** Francium (223)[a]	88 **Ra** Radium 226.0254[b]	89** **Ac** Actinium (227)[a]	104 **Rf** Rutherfordium 261	105 **Ha** Hahnium 262	106 **Unh** Unnilhexium 263	107 **Uns** Unnilseptium 262	108 **Uno** Unniloctium 265	109 **Une** Unnilhexium 266									

* Lanthanides

58 **Ce** Cerium 140.12	59 **Pr** Praseodymium 141.9077	60 **Nd** Neodymium 144.24	61 **Pm** Promethium (145)[a]	62 **Sm** Samarium 150.4	63 **Eu** Europium 151.96	64 **Gd** Gadolinium 157.25	65 **Tb** Terbium 158.9254	66 **Dy** Dysprosium 162.50	67 **Ho** Holmium 164.9304	68 **Er** Erbium 167.26	69 **Tm** Thulium 168.9342	70 **Yb** Ytterbium 173.04	71 **Lu** Lutetium 174.97

** Actinides

90 **Th** Thorium	91 **Pa** Protactinium	92 **U** Uranium	93 **Np** Neptunium	94 **Pu** Plutonium	95 **Am** Americium	96 **Cm** Curium	97 **Bk** Berkelium	98 **Cf** Californium	99 **Es** Einsteinium	100 **Fm** Fermium	101 **Md** Mendelevium	102 **No** Nobelium	103 **Lr** Lawrencium

[a] Mass number of most stable or best-known isotope.

[b] Mass of most commonly available long-lived isotope.